The Urban Book Series

The Urban Book Series is a resource for urban studies and geography research worldwide. It provides a unique and innovative resource for the latest developments in the field, nurturing a comprehensive and encompassing publication venue for urban studies, urban geography, planning and regional development.

The series publishes peer-reviewed volumes related to urbanization, sustainability, urban environments, sustainable urbanism, governance, globalization, urban and sustainable development, spatial and area studies, urban management, transport systems, urban infrastructure, urban dynamics, green cities and urban landscapes. It also invites research which documents urbanization processes and urban dynamics on a national, regional and local level, welcoming case studies, as well as comparative and applied research.

The series will appeal to urbanists, geographers, planners, engineers, architects, policy makers, and to all of those interested in a wide-ranging overview of contemporary urban studies and innovations in the field. It accepts monographs, edited volumes and textbooks.

Now Indexed by Scopus!

More information about this series at http://www.springer.com/series/14773

Feng Yang · Liang Chen

High-Rise Urban Form and Microclimate

Climate-Responsive Design for Asian Mega-Cities

Springer

Feng Yang
College of Architecture
and Urban Planning
Tongji University
Shanghai, China

Liang Chen
School of Geographic Sciences
East China Normal University
Shanghai, China

Funded by the National Natural Science Foundation of China (NSFC) Project (No.: 51678413)

ISSN 2365-757X ISSN 2365-7588 (electronic)
The Urban Book Series
ISBN 978-981-15-1713-6 ISBN 978-981-15-1714-3 (eBook)
https://doi.org/10.1007/978-981-15-1714-3

This Springer imprint is published by the registered company Springer Nature Singapore Pte Ltd.
The registered company address is: 152 Beach Road, #21-01/04 Gateway East, Singapore 189721, Singapore

Foreword

It is projected that some 70% of the world population will be living in urban areas. Tokyo is now the world's largest city with an agglomeration of 37 million inhabitants. This is followed by New Delhi with 29 million, Shanghai with 26 million, and Mexico City and São Paulo, each with around 22 million inhabitants. By 2030, UN predicts that there will be 43 megacities with more than 10 million inhabitants, most of them in developing regions.

Understanding the environmental qualities of these megacities and planning for more livable urban living spaces are important concerns for policy makers, developers, planners, and architects. A comfortable and clean living environment must be the starting point. The temperature one feels and the air one breaths immediately tells the body how conducive an environment is. The scientific components of thermal comfort are well established: air temperature, humidity, air speed, and radiation. When designing cities, planners, and architects must manipulate these components to achieve a diversity of conditions for the enjoyment of the city inhabitants.

This is where this book by Yang and Chen comes in. Both trained in Hong Kong—one of the world's highest density cities—they have approached the subject matter systematically. By using Shanghai as a case study, the explanations are more realistic and therefore practical to planners and designers. Chapter 1 outlines the scope of the subject matter to be dealt with. The term high-rise urban form is used to characterize megacities. Chapter 2 outlines the theoretical basis of the components of thermal comfort. Chapter 3 furthers with an appreciation of high-rise urbanism using Singapore, Hong Kong, and Shanghai as examples. Chapter 4 introduces the very important concept of urban heat island (UHI). The bigger and higher density the city, the more it will suffer from a hotter environment due to this urban heat island phenomenon. Urban design techniques to combat UHI are introduced in Chaps. 5–7. The book ends with a study of an urban area in Shanghai.

From theories to the study of a real case, the book unfolds a comprehensive understanding of the subject matter in a clear and systematic manner. As an architect myself, I particularly enjoy reading the last chapter on how knowledge may be put to practice. The book is a welcome addition to current literature.

I congratulate the authors for their hard work and revelations. I am sure it will be a useful reference to policy makers, developers, planners, and architects working in Shanghai. In a few years' time, I long to visit a few urban neighborhoods that are going to benefit from the authors' insights and teaching.

Hong Kong SAR, China
November 2019

Edward Ng
Yao Ling Sun Professor of Architecture
The Chinese University of Hong Kong
President, Passive and Low
Energy Architecture

Preface

The present book is a joint effort of two researchers from different background of disciplines (one Architecture and the other Computer Science), but both spent their Ph.D.-pursuing years in the City of High-Rise, Hong Kong, and chose urban form and microclimate as their these, and are both now stationed in Shanghai carrying out academic research, teaching, and practice. The collaboration with the above characteristics has produced the book that is inter-andmulti-disciplinary, is empirically based, focuses on Shanghai, and is oriented toward design application. The empirical approach is chosen not only because that the result may be more applicable in the investigated city and region, but also due to the belief that, any feasible solutions to urban problems can only be effectively explored in real-world cities. The book is the authors' effort in contributing to bridging the gap between architecture/urban design and urban climatology so that the two disciplines can better communicate, and knowledge and know-how regarding urban form verses microclimate can better inform climate-responsive design practice.

The authors would like to thank many people during the Hong Kong years for mentoring, collaborating and conversational exchanges that inspired, enlightened, and supported the authors' research work—Prof. Stephen Lau and Dr. Giridharan from the University of Hong Kong; Prof. Edward Ng from Chinese University of Hong Kong and his team, especially Dr. Yuan Chao from NUS Singapore. Thanks are due to M. Arch students from CAUP Tongji University, Ms. Wang Rui and Mr. Feng Yi-Peng for helping making some of the diagrams.

There seems to be two unclear or ambiguous boundaries in urban microclimate: one is physical-environmental, i.e., the multi-scale nature of urban meteorological phenomena and processes. To estimate the site-specific effect of urban design/building design at the microscale, we may employ careful boundary control in experiment design or develop multi-scale and coupled modeling technics in numerical simulation. The other one, overlooked in climate research but critical in application, is the socioeconomic one, i.e., the boundary between public domain, in which microclimate is an important merit, and other domains with various stakeholders: management, developer, owners, visitors, etc. Good microclimate in urban public space is like fresh air, everyone likes it but one will be held responsible for it.

However, for Asian megacities such as Shanghai, urban design guidelines specifically incorporating the aspect of urban microclimate is in urgent need, as the high-rise development pattern can lead to negative or even disastrous consequences. It is not until the outbreak of SARS in 2003 that the Hong Kong government started to realize the extent of negative impact the congested urban form can have, and initiated the Air Ventilation Assessment (AVA) effort and embarked on a decade of urban climate studies which lead to a series of now compulsory design and planning standards and guidelines. That lesson should be learnt.

It is time we take action.

Shanghai, China
October 2019

Feng Yang
Liang Chen

Contents

Part I

About the Authors

Feng Yang is currently an associate professor at the College of Architecture and Urban Planning, Tongji University, Shanghai, China. Dr. Yang's research focuses mainly on climate-responsive urban design, and passive and low-energy architecture. He has published extensively in peer-reviewed journals and also serves as manuscript reviewer for *Journal of Building and Environment*, *Energy and Buildings*, *Architecture Science Review*, etc. He is Principle Investigator of more than ten completed and ongoing research funds, including two China National Natural Science Funds (NNSF). He holds a B.Arch. degree from Hunan University, an M.Arch. from Tongji University, and a Ph.D. from the University of Hong Kong. He is visiting professor to School of Architecture, Rensselaer Polytechnic Institute, NY, USA, in 2016. He is an elected member of the Youth Committee member, China Green Building Council (CGBC). His research in building energy-efficiency design and technology won him the Third Prize of 2017 Science and Technology Award by the Shanghai Government.

Liang Chen is an currently associate professor at the School of Geographic Sciences, East China Normal University, Shanghai, China. He holds a bachelor's degree in Computer Science from Tsinghua University and a Ph.D. degree in Architecture from the Chinese University of Hong Kong. His research interest is in the spatial modeling and measurement of microscale urban climates, focusing on the influence of building morphology, land surface, and vegetation cover on human thermal comfort in a complex urban environment. He holds a number of research funds, including the National Natural Science Funds of China and the Natural Science Funds of Shanghai. He has published extensively in internationally renowned journals including *Building and Environment*, *Landscape and Urban Planning*, *Energy and Buildings*, *Architecture Science Review*, etc., and he is an active reviewer of these journals. He has participated in a number of research projects on urban climate modeling and application in Hong Kong, Japan, Sweden, Germany, and Singapore.

Acronyms and Abbreviations

BCR	Building cover ratio
CASBEE	Comprehensive assessment system for building environmental efficiency
CBD	Central business district
CCD	Cloud cover at day
CCN	Cloud cover at night
CGS	Central green station
DBT	Dry bulb temperature
EF	Ecological footprint
ET	Evapotranspiration
EW	Elevated walkway
FAD	Frontal area density
FAR	Floor area ratio
GCR	Green cover ratio
GPR	Green plot ratio
GR	Green ratio
GSA	Ground surface albedo
GSR	Global solar radiation
H/W	Aspect ratio
HPR	Hard pavement ratio
ITD	Inter-urban temperature differential
LCG	Lujiazui Central Green
LCZ	Local climate zone
LEED	Leadership in energy and environmental design
LEW	Lujiazui Elevated Walkway
LJZ	Lujiazui
MOHURD	Ministry of Housing and Urban-Rural Development of China
PET	Physiological equivalent temperature
PHS	Proximity to heat sink
RH	Relative humidity

SRS	Solar radiation on site
SVF	Sky view factor
T_{mrt}	Mean radiant temperature
Ta	Air temperature
Tg	Globe temperature
TMY	Typical meteorological year
TSF	Total site factor
TSV	Thermal sensation vote
TVF	Tree view factor
UBL	Urban boundary layer
UCL	Urban canopy layer
UHI	Urban heat island
VTD	Vehicle traffic density
WD	Wind direction
WV	Wind velocity
WVD	Wind velocity at day
WVN	Wind velocity at night
WVR	Wind velocity ratio

Part I

Chapter 1
Introduction: High-Rise Urban Form and Microclimate

Abstract This introductory chapter serves to set the scene of the whole book. Many megacities emerged in Asian countries accompanied with the recent unprecedented urbanization process. Many researches indicate that these megacities are prone to negative consequences in urban microclimate. These are indicated by elevated urban heat island effect, stagnant urban air, and deteriorated urban outdoor comfort. The resultant air pollution, impaired human health, and extra costs in air-conditioning energy bills will make these cities less environmentally and socially sustainable. While many of these cities adopt a high-rise development pattern, it is unclear that the resultant high-rise urban form can be a cure for, or were the curse of the problems mentioned above, particularly when it comes to the district/neighborhood scales. It is well-established that urban form has an impact on the microclimate, and climate-responsive or bioclimatic building and urban design strives to help create low energy, natural, and comfortable habitable urban spaces through manipulating urban form (urban geometry, building materials, and land cover). Shanghai is chosen as the case for study, and the urban and climatic context of China's largest city are briefly introduced. The chapter is closed by an outline of the following chapters.

1.1 The Issues Arising from Rapid Urbanization

Asia and other economically less-developed regions have been experiencing an accelerating pace in urbanization, and this process shows no sign of retreat over the next decade. While half of the world's population had become urban dwellers by 2006, the number of Chinese citizens living in urban areas reached 50% of the population in the country in 2011. In 2010, the World Exposition was hosted in Shanghai, the largest city in China. Not coincidentally, the theme of this World Expo is exploring the prospect of future urban life, which is clearly shown in the official slogan "*Better city, better life*." It epitomizes global concerns regarding improving urban life in the face of all of the challenges posed by rapid urbanization, especially in developing countries.

Urbanization results in a concentration of various human activities; this concentration often causes heavy interferences in the operation of natural systems (Oke

F. Yang and L. Chen, *High-Rise Urban Form and Microclimate*,
The Urban Book Series, https://doi.org/10.1007/978-981-15-1714-3_1

1987). The problem is particularly pressing in densely populated "megacities (population larger than 10 million)." Till 2016, Asia has eight of the top ten largest cities of the world (Demographia World Atlas 2016). In these cities, economic growth has usually been given the highest priority, thereby causing various loads on the urban environment, which has seriously affected urban living standards.

Urbanization has a profound effect on the meso, local- and microscale climates in the urban area. The resulted urban climate has significant impacts on urban ecological system and human habitation. Regarding the thermal environment, the urban heat island (UHI) effect may be the most studied subject of urban climate research. Indeed, it epitomizes the synthetic effects of an urbanized area on its climate, using one single index of temperature. An UHI is the thermal gradient difference as measured between a developed area and an undeveloped area (USGBC 2005). A potential favorable phenomenon in colder climates, UHI increases discomfort and building cooling energy demand in warmer climates (Schiller and Evans 2000). In fact, UHIs can give rise to or exacerbate many problems, such as changes in wildlife and vegetation ecosystems, increasing numbers of stroke patients, expansion of infectious diseases, in addition to increasing demand for building cooling energy consumption, all of which can threaten the living environment and health of urban inhabitants. UHI is a multi-scale phenomenon, and researchers have long recognized that the UHI effect is partially due to choices in construction materials, greenery, and urban design (Santamouris 2001).

Urban form helps shape the urban microclimate, an important but often neglected aspect of quality of urban outdoor spaces used by city dwellers. Thermally comfortable outdoor spaces may attract more pedestrians and participants, catalyzing diverse public activities and thus contributing to more lively and socially sustainable urban space (Gehl 2010). The thermal and aerodynamic properties of urban structure and fabric influence the energy and mass transfer processes that characterize the urban microclimate, dynamic and diverse in solar thermal, airflow and temperature distribution below the average roof level of urban buildings, in accordance with the diversity in urban form (Erell et al. 2011). On the other hand, human thermal comfort is influenced by environmental factors of air velocity, humidity, and radiant temperature in addition to air temperature. Compared to an indoor environment, thermal comfort assessment in an outdoor setting is more dynamic and context-specific (Steemers and Steane 2004). In dense urban districts commonly found in emerging contemporary Asian megacities, the high-rise urban form dramatically changes the pedestrian-level exposure to sun, light, and airflow. How is the change related to the comfort and livability of urban spaces, and what can urban planning and design do in making it a better place to live in?

Therefore, combined with hot–humid climates, the urbanization in Asian large cities poses critical challenges as well as valuable opportunities for urban planners and architects interested in climate-responsive design. The objective of this book is to respond to these challenges and opportunities. By a series of application-oriented research, it provides readers with knowledge and technics in climate-responsive design that is pertinent to the unique urban and climatic context of Asian megacities.

1.2 Adverse Effects of Urbanization on Urban Climate

1.2.1 Air Pollution and Smog

Air pollution is a serious issue accompanied with rapid urbanization in South-East Asia regions and particularly in the megacity clusters. According to a recent study on air pollution records of 336 Chinese cities, the concentrations of $PM_{2.5}$, PM_{10}, SO_2, NO_2, and CO are highest in winter, while the concentration of ozone (O_3) is highest in summer (Li et al. 2019). Heavy industrial and traffic vehicle emissions, stagnation due to the wall effect of densely packed tall buildings, and adverse meteorological conditions all contribute to worsening air quality. Therefore, from the perspective of urban design, improving urban ventilation can be an effective strategy in purging urban air of pollutants, especially those released by vehicles (Hang et al. 2012). The "breathability" is thus an important indicator of urban form in assessing the climatic impact of high-rise cities (Ng et al. 2011).

Unlike the O_3 layer in the stratosphere, which protects life on earth by absorbing ultraviolet light from the sun, ground-level O_3 has detrimental effects on human health and the environment. The negative physical effects on humans are mainly associated with human respiratory system diseases and disorders (Kimberly and Finster 1999). The O_3 intensity increases with an increase in ambient air temperature and solar radiation (Chow and Su 1994). For instance, a numerical study indicates that a UHI reduction of 1–2 °C in three US cities would yield a reduction in O_3 concentration of 3.5–7% as compared to peak values (Taha et al. 2000). Therefore, lowering summertime UHI can potentially contribute to air quality improvement.

1.2.2 Thermal Discomfort and Impaired Human Health

Human thermal comfort status is a balance among several factors that can be grouped into the following three categories: environmental factors, including air temperature, humidity, radiation, and air movement; personal factors, including insulation level (expressed in clothing units) and metabolic rate; and contributing factors, including food and drink, acclimatization, body shape, subcutaneous fat, age, and gender (Fanger 1970; Szokolay 2008). Indoor thermal comfort has been extensively studied, and some evaluation standards have been established, e.g., the ANSI/ASHRAE Standard 55. Quantifying thermal comfort under *outdoor* conditions, however, needs to take into account the complexity of the outdoor wind and radiation environment as well as differences in the nature of outdoor human activities. Urban form and its outdoor thermal comfort impact have been a very active interdisciplinary research field in architecture, urban planning, and climatology in recent years, and plenty of design methods and tools have been developed to better inform climate-responsive design process. One of these tools, called *thermal atlas* that is developed by the authors, is introduced in Chap. 8.

UHIs can affect human health by contributing to general discomfort or physical disorders and diseases, such as respiratory difficulties and heat stroke. A strong UHI effect that accompanies heat waves can contribute to significant increases in heat-related mortality. Its negative impacts are particularly detrimental to certain sensitive groups, such as children and the elderly. Worldwide, one such heat wave in the Midwest of the USA caused more than 1000 deaths (Taha et al. 2004). The 2003 heat wave that swept through Europe resulted in an increase in the death toll in the order of 22,000–50,000 (Schär et al. 2004). Shanghai is particularly prone to extreme heat. Recorded heat waves from the past several decades have led to elevated levels of human mortality rates. For instance, an intense heat wave in 1998 resulted in a large number of deaths (Tan et al. 2007). Moreover, the population of Shanghai is aging significantly on average. By definition, a senior urban dweller society is the one with its senior city dwellers (i.e., 65 years and older) population reaching 7% of total population. Senior city dwellers in Shanghai accounted for 21.8% of the city's overall (registered 户籍) population by the end of 2017 (Shanghai Research Center on Aging 2018). This percentage is the highest among the China's first-tier cities including Beijing, Shanghai, Shenzhen, and Guangzhou. It is also one of the highest among big cities in the world. This part of the Shanghai population would be particularly vulnerable to excessive heat events. Thus, the UHI effect as a contributing factor to heat disasters deserves an in-depth investigation.

1.2.3 Increasing Building Cooling Energy Consumption

Building operational energy consumption accounts for a considerable part of the overall energy use by society. Summertime UHI increases energy demand in terms of indoor cooling and adds a burden to peak loads. In the USA, the peak urban electric demand increases 1.5–2% for every 1 °F (0.6 °C) increase in summertime air temperature (EPA 2002). With ambient temperatures continuing to increase in downtown areas, 5–10% of a community-wide electrical demand is used to compensate for the UHI effect (Akbari and Konopacki 2005). The increase in energy bills as associated with UHIs is quite large. In the 1990s, UHIs raised air-conditioning demand by about 10 GW in the USA, which ultimately cost several million dollars per hour and a billion dollars annually (Rosenfeld et al. 1995). A study in Bangkok also revealed that an increase of 1 °C in ambient air temperature results in an increase of 6.8% in electricity consumption (Wangpattarapong et al. 2008).

The number of household-owned air conditioner units in the urban area of Shanghai is increasing at a very fast pace. Shanghai had 205 air conditioners per 100 urban households in 2016 as compared to only 136 in 2003 (Shanghai Municipal Statistics Bureau 2008). This increasing number is closely related to summertime cooling energy use. The energy used for air-conditioning in summer months was estimated to account for as much as 40% of the total energy demand in Shanghai (Glicksman and Lin 2006). Mitigating UHI effect could thus contribute to the significant reduction in energy usage for housing cooling.

1.3 High-Rise Urban Form and Environmental Sustainability

Urban form is defined by Lynch as "the spatial pattern of the large, inert, permanent physical elements in a city (Lynch 1984)." It is believed by many that a sustainable urbanization needs careful consideration on its urban form with respect to compactness, density, mixed land use, diversity, passive design, etc. (Jones and Storey 2017). In the fields of urban environmental physics and urban climatology, studies on urban form consider specifically the urban geometry (structure), fabric (material), and land cover. As these aspects, in addition to urban metabolism, largely determine the energy and mass transfer and balance of the urban built environment (Oke et al. 2017).

The word "sustainable" is derived from the Latin verb *sustinere*, which describes relationships, processes, or states that can be maintained indefinitely or for a long period of time (Tzonis 2006). The term *sustainable development* has gained popularity worldwide since the release of the report entitled *Our Common Future* by the World Commission on Environment and Development (WCED) in 1987. It is defined as "development that meets the needs of the present without compromising the ability of future generations to meet their own needs (WCED 1987)." One of the two key concepts is "the idea of limitations imposed by the state of technology and social organization on the environment's ability to meet present and future needs (WCED 1987)." The other concept is that of "needs," specifically the essential needs of the world's poor. Both concepts are interrelated from the standpoint that more and more pressure is being exerted by human society on the natural environment with the continuing growth of the world's population. This pressure is particularly intense from populous developing countries in which the people require a higher standard of living than before, based on the socioeconomic progress that they have achieved. It is widely accepted that there are three pillars of sustainability, i.e., economic, social, and environmental. The interest of this book is on the environmental aspect.

Generally, the high-rise building typology is not usually considered environmentally sustainable, due to its intensive use of energy, material, and other resources during its whole life cycle, although it has merits, such as saving land resources and transportation costs (Gonçalves and Umakoshi 2010). When assessing the environmental sustainability of high-rise buildings, often a target of "doing less harm" instead of "doing more good" is set. A number of tall buildings featured with sustainable environmental design can be found in different geographical locations with various climates, for example, the Commerzbank in Frankfurt, Germany, designed by Foster and Partners, Menara Mesiniaga in Subang Jaya, Malaysia, by Ken Yeang, Deutsche Messe AG in Hannover, Germany, by Thomas Herzog, KfW Westarkade also in Frankfurt, Germany, by Sauerbruch Hutton, etc. Invariably, they focus on individual tall buildings, rather than a group of buildings. By contrast, the sustainable environmental urban design of high-rise buildings is much less studied. Research at larger scales is in an urgent need, as high-rise urban form is changing the skylines of Asian megacities; whereas, a number of environmental issues (urban heat islands, air pollution, thermal comfort, etc.) are considered related to high-rise, high-density urban development.

The high-rise urban form found in Asian megacities can be traced back to the Radiant City advocated by Le Corbusier nearly 100 years ago, which has had profound influences on modern urban planning. Building into the sky may be rationalized in intensive utilization of high-value urban land, but tall and super-tall buildings can also be perfect symbols of the economic achievement of a city and even a country. In the residential sector, living high means more daylight and cool breeze while less humidity (and maybe more clean air, fewer mosquitoes, and thus overall improved sanitary status). These are probably the understated reasons why high-rise living is embraced by urban dwellers in large Asian cities which have hot–humid climates (Kubota et al. 2008). But when it comes to the spaces in between buildings, the environmental quality is often much less concerned probably due to less economic interest. However, it must be recognized that the quality of urban outdoor spaces is equally important for an environmental and socially sustainable city, as it is for the common good of urban dwellers.

Therefore, people who plan, design, and manage Asian megacities must seek and maintain environmentally sustainable development patterns during the socioeconomic progression. The essence of sustainable urban living lies in improving living standards while reducing impact on the natural environment. From an urban design perspective, we must seek approaches to improve both the indoor and outdoor urban physical environment, in order to enhance public health and promote more outdoor activities and to reduce the ongoing expansion of operational energy consumption and thus non-renewable energy use and air pollution, all of which calls for a more environmentally conscious planning and design paradigm.

This book focuses on modifying the urban microclimate through climate-responsive urban design approaches, based on the recognition that urban areas with low climatic qualities generally contribute to exacerbated air pollution levels, more energy used for conditioning buildings, and less comfortable outdoor spaces that are not user-friendly, thereby threatening the prospect of sustainable urban living.

1.4 Design with Climate: The Bioclimatic Design Approach

Research on climate responsiveness in urban planning and building design should be seen as a subset of broader environmental design strategies. Specifically, this area of study is closely related to the bioclimatic design approach. In his seminal work *Design with Climate: Bioclimatic approach to architectural regionalism* published in the 1960s, Olgyay proposed a systematic design and research methodology to adapt the built environment to its climate and its people by considering the following four aspects: climatology, biology, technology, and architecture (Olgyay 1963). In essence, a bioclimatic approach starts with a summary and analysis of local meteorological data and its impact on human sensations. Technological solutions are then sought to balance the built environment in response to the aforementioned impacts, either adverse or advantageous. The output must be finalized and presented in a form with architectural integrity. Inspired by Olgyay's work, Yeang developed a research and design approach called *bioclimatic skyscrapers* for designing climate-sensitive tall buildings in the tropics (Yeang 1994). At the local and urban scale, the bioclimatic design is even more important and useful in Asian subtropical cities than in cooler mid- to high-latitude cities. By carefully considering the building density, sizes, and distances of building and grouping of buildings and proper mixing of urban vegetation with respect to the specific climatic features, outdoor comfort and air quality may be improved by efficient shading, ventilation, and evaporative cooling. This has been well-summarized as sets of design guidelines by Givoni in his book *Climate Considerations in Building and Urban Design* (Givoni 1998).

Landscape design and, in particular, vegetation are an indispensable part of bioclimatic design, as urban greenery offers climatic, ecological, social/psychological, and economical benefits (Givoni 1998). Yeang proposed several ways to vertically combine vegetation with high-rise built form, including juxtaposing, intermixing, and integrating (Yeang 1994). The radiative, aerodynamic, thermal, moisture-related, and structural properties of urban greenery distinguish it from other urban surface materials. The resulting ability to provide coolness and shade for urban buildings or open spaces makes it a flexible tool for environmental design (Oke et al. 1989). The effective environmental cooling performance of urban greenery mainly relies on evapotranspiration and canopy shading; the former lowers air temperatures by transferring sensible heat into latent heat fluxes, while the latter prevents temperatures under the vegetation canopies from rising by intercepting solar radiation. However, within a high-rise built environment, these biological and physical processes become much more complicated for urban greenery as compared to its rural counterparts, because of the influences from dense buildings, canyon-like street geometry, and impervious construction materials. Landscape design should not be isolated from building planning and design. The parameters related to landscape, planning, and building design should be considered and manipulated in a collaborative manner because all of these parameters that interact with each other to shape the thermal environment.

The realm of green building or GB is associated with the bioclimatic building in the aspect of responding to the climate (or passive design) so as to minimize energy use by active mechanical systems for a satisfactory comfort level. The green building however is more comprehensive, considering the energy and resource consumed, environmental quality achieved for the habitable space under assessment, as well as environmental impact from local up to the global scale through the life cycle of buildings (MOHURD (Ministry of Housing and Urban-Rural Development) 2019). The GB rating tools are closely tied with nation's socioeconomic status and heavily influenced by political and market forces. Nevertheless, it is worth mentioning that the bioclimatic design efforts to mitigate UHI and improve outdoor thermal comfort are encouraged by a number of GB rating tools developed in different countries. For instance, the green building rating system of the USA, known as Leadership in Energy and Environmental Design (LEED), offers credits for efforts made to reduce UHI by applying on-site shading, vegetation, or roof materials with high solar reflectance values (USGBC 2005). In the Evaluation Standard for Green Buildings promulgated by the Ministry of Housing and Urban-Rural Development of China (MOHURD), credits are offered to countermeasures of UHI including provision of shading, cool pavement, and green roof (MOHURD (Ministry of Housing and Urban-Rural Development) 2019). The Japan's green building evaluation system known as Comprehensive Assessment System for Building Environmental Efficiency (CASBEE) developed a specific sub-manual, i.e., CASBEE-Heat Island relaxation (CASBEE-HI) (Japan Sustainable Building Consortium 2005). When evaluating an amelioration approach, five aspects, i.e., site ventilation, shading, ground surface cover, building cladding materials, and anthropogenic heat, are assessed separately, and the results are then integrated with various weighting factors.

1.5 Shanghai: Urban and Climate Context

This book bases the study in the city of Shanghai, China. According to statistics, Asia has eight of the top ten megacities in the world. Shanghai is ranked the eighth, with a population of 24.2 million in 2017 (Shanghai Statistical Yearbook). Shanghai represents current and future megacity development in China or even other Asian developing countries, due to a number of reasons. First, Shanghai is the largest city in China, the world's most populous country. Second, Shanghai is situated at the heart area of the Yangtzi River Delta Region, the economically most developed region in China, and regarded as the head city of the Yangtzi River Delta Urban Agglomerations, comprised 26 cities and contributing to nearly 20% of the national GDP (data of 2014). And third, according to China's building energy conservation code, Shanghai is located in the so-called hot-summer cold-winter building climate zone, occupying one-fifth of China's total land area. The economic activities in this climate zone produce about half of the national GDP (MOHURD (Ministry of Housing and Urban-Rural Development) 1994). Therefore, studying the relationship of urban form and its microclimate in Shanghai can be a very useful reference to other Chinese cities in the similar geographical, politic–economic, and cultural contexts.

1.5.1 High-Rise Urban Form in Shanghai

By definition, high-rise buildings in China comprise residential buildings higher than 27 m and non-residential buildings higher than 24 m and are equipped with lift(s) (MOHURD (Ministry of Housing and Urban-Rural Development) 2014). High-rise buildings as a building type have a long history in Shanghai. As early as in the 1920s, the first tall apartment building was built in the then French Concession area, now Xuhui District in Shanghai. From 1928 to 1948, 35 high-rise apartment buildings were built in the districts of Xuhui, LuWan, and JingAn. The tallest one of these buildings is 20 floors high (Li 2005). After the national economic reform policy in the late 1970s, and as a result of limited land, soaring land prices, and the ever-increasing population, high-rise buildings have been increasingly constructed and have gradually dominated the new development market in inner-city Shanghai (Fig. 1.1). Data from the Shanghai Statistics Bureau showed that up to the year 2017, there is a total of 44,395 high-rise buildings in the city, which adds up to a total floor area of 436.48 million square meters (Fig. 1.2). Noted that Pu Dong ND and Min Hang are larger districts, and thus, have more tall buildings.

Compared with low- to mid-rise housing, high-rise residential developments are considered to offer lower building cover ratio and improved wind environments, thereby making them more suitable for inner-city areas. Before 2000, the housing stock in Shanghai mainly comprises high-density and low-rise "Li-Long" row houses and mid-rise "workers" village. Low-rise (i.e., 2–3 stores) high-density "Li-Long" housing is a traditional housing type developed before 1949 to address the needs for mass living. Due to its high building cover ratio (i.e., usually exceeds 60% and sometimes peaks at 80%), high thermal-absorptive dark roofing materials, and extremely low green cover, solar absorption and re-radiation are enhanced, while wind tends to be stagnated in the Li (main lane) and Long (side alley) area of this row house development. It is no wonder that a study in 2002 on the contribution of land use to the UHI effect in Shanghai found that the residential land use contributes to the formation of the strongest UHIs, followed by transportation and industrial land use (Ding et al. 2002a). The mid-rise (4–6 stores) "workers" village in Shanghai was mainly built

Fig. 1.1 People's Square of Shanghai as seen today. The white arrow points to the Park Hotel, the tallest building in Shanghai (and in China), until the early 1980s (*Photo credit* Minggang Yin, used with permission)

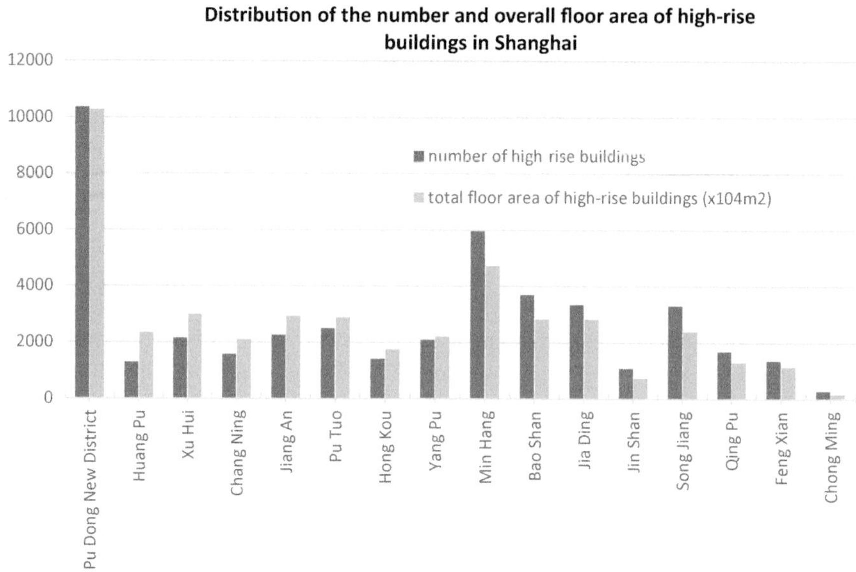

Fig. 1.2 Distribution of the number and overall floor area of high-rise apartment buildings in inner-city Shanghai in 2017

during the period from 1949 to 1990s. Nowadays, most newly developed housing projects in Shanghai consist of high-rise buildings (Fig. 1.3).

Compared to the traditional "urban-suburban-rural" land-use pattern, the results of urbanization in Shanghai and other Chinese cities appear to be more diversified, dispersed, and decentralized. Land-use control is often less strictly implemented, even with comprehensive urban planning ordinances available at hand. Newly developed high-rise building clusters are often just mixed into the underdeveloped suburban or even rural landscape. This intense mixture of land use and function makes it difficult to detect the urban edge. This situation seems to complicate the UHI research because the determination of a "rural" site in contrast to its "urban" counterpart can be difficult. It would be more meaningful to study the intra-urban difference in local to microclimate and its relationship to urban form. The local climate zone scheme (LCZ) proposed by Stewart and Oke (2012) could serve this need. It could also potentially facilitate a comparison of UHI studies on various cities under different climate. The LCZ schemes are discussed in greater detail in Chaps. 3 and 4.

For high-rise building clusters, the regulatory framework for the spacing control between tall buildings is critical in determining built form and density. In Shanghai, this is mainly based on the compulsory requirement of sunlight access to the indoors with due consideration for fireproofing and other planning needs (Table 1.1). Among the above factors influencing site layouts, the availability of sunlight to indoor areas is the most important. Based on this premise, the floor area ratio (FAR) could be maximized. Additionally, due to the requirement for setback from redlines, the distance between residential buildings on the opposite sides of an east–west running

Fig. 1.3 Urban Shanghai, a juxtaposition of low-rise Li-Long, mid-rise workers' village, and contemporary high-rise buildings (*Photo credit* Minggang Yin, used with permission)

Table 1.1 Spacing control of high-rise residential buildings in inner-city Shanghai

Requirement	Spacing control
Insolation duration/natural lighting	1 h or more duration of direct sunlit during the effective period (i.e., 9 a.m.–3 p.m.) for major living spaces of every unit on winter solstice
Fireproofing	High-rise to high-rise: $\geq$ 13 m; High-rise to low-rise: $\geq$ 9 m; Low-rise to low-rise: $\geq$ 6 m
Privacy and ventilation	Spacing of lateral walls: At minimum, 1/4 of the height of the higher one between two lateral walls; AND at minimum, 24 m in N–S or 18 m in E–W direction between lateral walls, whichever is larger
Setback	Setback from red lines of roads: $\geq$ 5 m for high rises lower than 60 and $\geq$8 m for high rises higher than 60 m

Source Multiple national building codes in China

road that separates two blocks is usually larger than the space between two buildings within one block. This could explain why the higher buildings are almost always placed on the northern side of a block. As a result, the layout of a residential block in Shanghai often appears to be "loose inside and dense outside, high in the north and low in the south (Li 2005)."

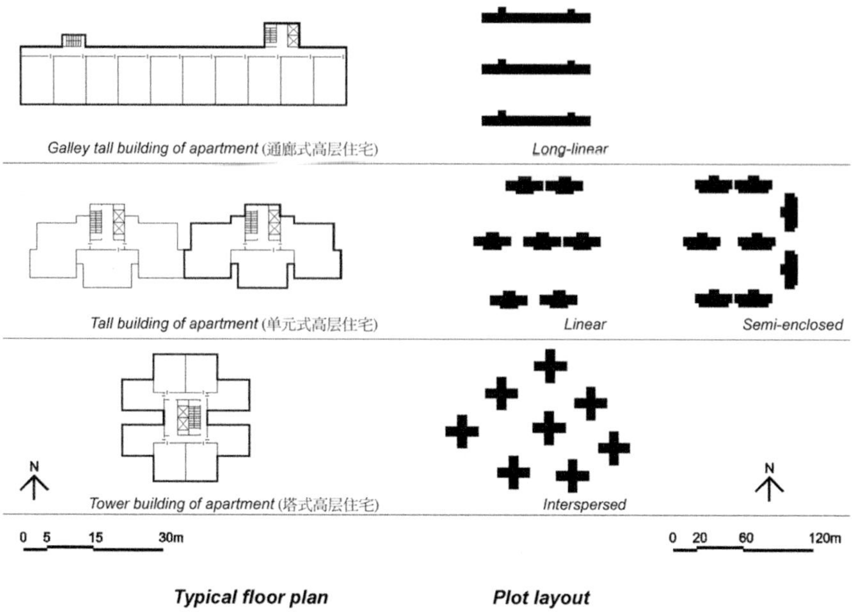

Fig. 1.4 Diagram of typical floor plans and plot layouts of high-rise residential developments

In general, most typical floor plans for high-rise residential buildings in today's Shanghai can be generally categorized as gallery tall apartment buildings, tower apartment buildings, and tall unit apartment buildings[1] (Fig. 1.4).

The *gallery tall building* features a long galley or walkway, which is normally located on the northern side of the floor plan and provides access from every housing unit to the vertical service core(s). The length of the typical floor plan can be as much as 60 m and accommodate up to 20 units. This floor type offers a very high saleable FAR but also provides the lowest privacy level as compared to the other types. The design and construction activities related to gallery tall buildings were at the highest in the 1970s in Shanghai, but this type of construction has largely been retired from the housing market after the 1990s. A site plan with this kind of tall building usually features a long-linear pattern.

The *tall unit apartment building* is an improved floor type that is based upon the previous type. A tall apartment building resembles the gallery type, but the galley is dramatically shortened so that each floor typically accommodates only two to four units. The privacy is thus significantly enhanced. Unlike the tower type, the major living space in every unit in this building type faces south, and efforts are made to enable every unit to enjoy some form of cross-ventilation. When applied to a site, two to four of the typical floor plans with separated vertical service cores and sealed lateral walls can be readily connected and combined into a longer slab in the longitudinal direction, thereby using the plot more efficiently. With the above merits,

[1]Source: Design code for residential buildings, China GB 50096-2011.

this kind of floor plan gradually gained dominance in Shanghai after the 1990s in conjunction with the prosperity of the private housing market. The plot layout for a site that uses this floor plan can be classified as either a linear or a semi-enclosed plot depending on the actual organization of the typical floor plans.

The *tower apartment building* uses a typical floor plan that locates the service core in the center with units surrounding it. A typical floor normally accommodates 6–8 units. This floor type is much smaller in size than the gallery type, so it is more appropriately suited to smaller sites with irregular shapes. Only a minority of units faces south, which is the most favorable orientation in the geographical and climatic context of Shanghai. Therefore, this is a major drawback for this floor type. The tower apartment buildings were mostly built in the 1990s in Shanghai. A site that includes this type normally exhibits an interspersed layout.

Of the four plot layouts generated by combinations of the three typical floor plans, i.e., *long-linear, interspersed, linear,* and *semi-enclosed,* each has its pros and cons from the occupants' viewpoint. In the context of Shanghai, the *linear* layout has become the mainstream choice in the apartment market. This is because of a populous belief that the linear type is the best fit for the climatic and geographic context of Shanghai. Since the subtropical climate and the strong influence from seasonal monsoons result in a hot, humid summer and a cold, humid winter in Shanghai, only rooms with southern exposure offer residents low-angle solar access in winters, shelter from cold-winter winds from the northwest, and pleasing breezes in the summers. The *interspersed* layout is flexible in terms of planning and offers the ability to achieve the largest saleable floor area from a given plot. As previously discussed, the orientation of its units is the major drawback. The *long-linear* layout formed with gallery tall buildings has not been used since the 1990s due to several serious drawbacks. In addition to the lack of privacy, bulky long slab buildings also inhibit summertime site ventilation and cast extensive shadows on areas on the northern side of the buildings, which creates unfavorable conditions in terms of both indoor and outdoor thermal comfort and activities.

However, the preference for the linear layout is based primarily on the resulting indoor thermal comfort levels. From a broader perspective, the methods by which different approaches in organizing building mass influence the on-site *outdoor* thermal environment have not been studied sufficiently. A better understanding of these influences can help in terms of design decision-making with the aim to improve outdoor thermal comfort and mitigate summertime UHIs. Empirical research on outdoor thermal environments related to housing from this perspective is scarce. Findings derived from analyses based on real on-site data are thus of high value because these findings reflect the complex, real-world situation and can complement, modify, or correct findings from numerical studies. Therefore, based on the above discussion, investigating the outdoor thermal environment of inner-city high-rise housing developments in Shanghai is imperative in terms of improving thermal comfort and reducing housing energy consumption.

1.5.2 The Climate of Shanghai

The city of Shanghai is located at the following geographical coordinates: latitude 31.23 °N and longitude 121.47 °E. The climate is classified as subtropical climate. In the summer months between June and September, the weather is usually hot and humid.

The chart below shows monthly temperature and precipitation for a typical meteorological year (TMY) in Shanghai as derived from data from 1971 to 2000 (Fig. 1.5). The highest monthly mean daily air temperature and mean daily maximum air temperature are 27.8 and 31.6 °C, respectively; both occur in July. A major portion of the annual precipitation occurs in the period from June to September. The highest amount of monthly rainfall occurs in June and September at about 150 mm. Although the amount of rainfall in July and August, (i.e., about 130 mm), is lower than the peak value, it is still significantly higher than the average rainfall in other months.

The temperature of Shanghai has been increasing in recent years. Figure 1.6 shows that the monthly mean daily temperatures over the past five years (2003–2007) has increased by about 1.8–2.9 °C as compared to the data from 1971 to 2000.

The increase in temperature is presumably not caused by an increase in solar radiation because the mean sunshine duration has been markedly reduced over the past five years, specifically in the hot months of July and August (Fig. 1.7). In addition to the macro factors that affect temperature elevation (e.g., global and regional warming), this increase in temperature is likely caused in large part by urban modification, which has changed urban surface materials and geometry, increased anthropogenic heat emission, and amplified air pollution levels.

The UHI effect has been a historical reality in Shanghai. Research by Chow shows that over the last century, the UHI effect in the urban areas of Shanghai is large and has been enhanced with time. In addition, over the last few decades, the

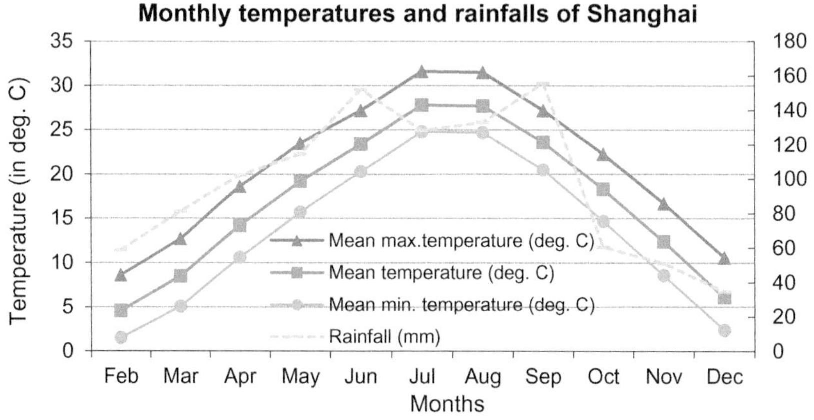

Fig. 1.5 Monthly mean temperatures and rainfalls of Shanghai (1971–2000)

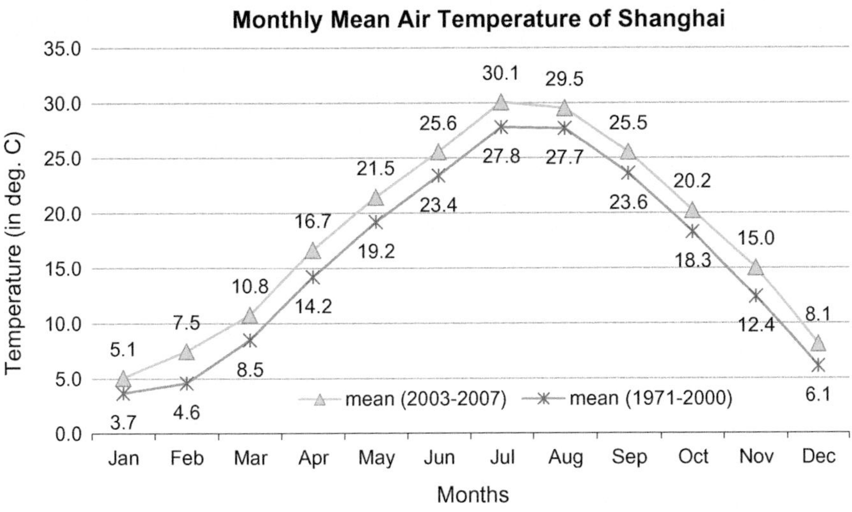

Fig. 1.6 Comparison of monthly mean temperatures: (1971–2000) versus (2003–2007)

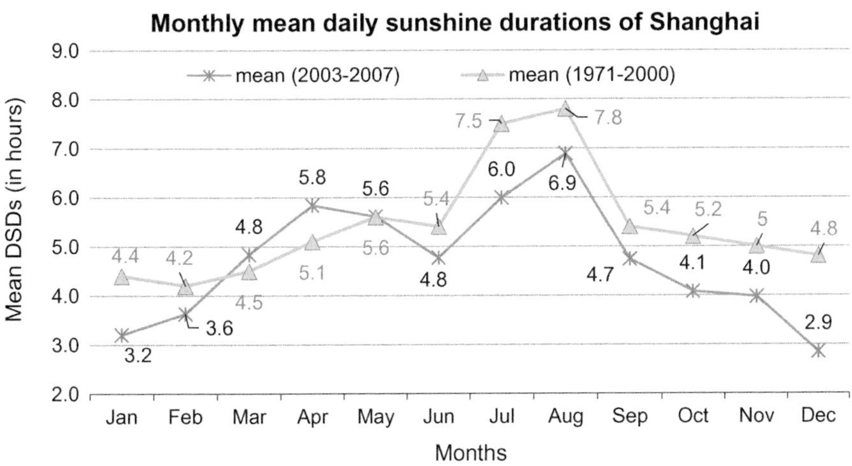

Fig. 1.7 Monthly mean daily sunshine durations (DSD) of Shanghai

inner city of Shanghai has suffered from adverse climatic changes, such as lower wind speeds and humidity, more fog, and an increased number of overcast days, due to rapid urbanization (Chow 1992). Urban surface temperature of Shanghai is found to be closely related to nocturnal air temperature UHI effect. Vegetation density influences urban surface temperature, and thus, the UHI in urban areas of Shanghai. Chow attributed the formation of the Shanghai UHI in large part to the urban–rural differentiation in thermal properties of the land cover (Chow et al. 1994).

1.5.3 Green Spaces in Shanghai

With increasing awareness of the benefits of green space on urban environments as well as city image, the Shanghai administration has implemented a greening campaign as an attempt to turn Shanghai into a "garden city." The focus is to enlarge public green space by constructing new parks and green lands in urban core areas. According to the data from the Shanghai City Development Research and Information Center (SCDIC), by 2003, the green area ratio, green coverage ratio,[2] and public green area per capita of the metropolis of Shanghai (including all the eighteen districts) had reached 34.51%, 35.72%, and 9.20 m^2, respectively. The resulting environmental effect has been encouraging. An urban survey using infrared images indicated that greenery-covered areas, such as Yan Zhong Greenland, have a lower surface temperature than their surrounding areas (SCDIC 2003). Another study on the cooling effect of Yan Zhong Greenland found that in summer, this area reduces the monthly mean daily maximum air temperature by 0.6 °C, and within three years of the launch of the greening campaign, the heat island intensity within the central area of the city was reportedly reduced by 0.2–0.4 °C (Ding et al. 2002b).

The effort abovementioned by the government is only focused on public green spaces. Residential green spaces, which represent a larger portion of the overall green space in the inner city than the public green space, deserve an appropriate amount of attention regarding the potential thermal benefits. Residential green space is defined as the space covered with vegetation within a residential development (MOHURD (Ministry of Housing and Urban-Rural Development) 2016). Residential green space constitutes an important part of urban greenery in the inner city of Shanghai (Table 1.2).

The data shows that the overall size of green space in the inner city is very small at only 12.28 km^2. Nevertheless, the residential green spaces are major components of this green space system. Indeed, these spaces represent the second largest category by area and comprise 23% of the overall green space; they are preceded only by natural vegetation (i.e., 32.21%) (Che 2004). Considering the size, if organized and designed properly, residential green spaces could play a significant role in improving the outdoor thermal environment within and even beyond residential developments.

[2]Green area ratio (绿地率) is the metric to measure the land use of green space in residential development and urban planning. The green spaces counted in green ratio calculation should meet the minimum depth of substrate soil of 3000 mm. Green coverage ratio (绿化覆盖率) means the ratio of ground-projected vegetation canopy area (tree, shrub, grass, etc.) to the overall land area. It has no restrictions on the depth of soil. All kinds of greening (e.g., rooftop garden) can be counted in GCR calculation.

Table 1.2 Composition of five types of green spaces in inner-city Shanghai [48]

Category	Area (in Km^2)	Percentage (in %)
Public green space	1.80	14.64
Residential green space	2.82	22.99
Auxiliary green space	2.52	20.53
Agriculture–forest vegetation	1.18	9.63
Natural vegetation	3.96	32.21
Total	12.28	100

1.6 Scope and Structure of the Book

Studies on causal relationship between urban form and climate have been carried out in many countries around the globe. From the perspective of urban and building design, a general consensus suggests that the following can be used to modify urban climates: increasing effective vegetative density, using appropriate building materials, and manipulating urban geometrical parameters. The thermal performance of urban geometry, fabric and greenery parameters, and their interplay are the focus of this book.

For climates with hot–humid summers and cold winters, the priority in terms of design should be given to solving heat stress problems in the summertime (Givoni 1998). This is why urban and building design in this climate zone should adopt a climate-responsive design approach aiming at lowering summertime air temperatures and improving the overall thermal conditions. While the indoor thermal environment and comfort issue have been intensively studied worldwide, outdoor thermal conditions and the impacts on air quality, human health and comfort, and building energy use have been relatively ignored, especially in large cities in developing countries. In those high-density megacities in Asia such as Shanghai, the ability to obtain sunshine, wind, and fresh air in a dense man-made environment is highly valued and yet becoming increasingly difficult. Additionally, any attempt to enjoy these assets would be economically and environmentally unsustainable in the absence of a supportive outdoor environment.

Findings of studies from other regions related to the relevant topic are difficult to apply directly to Asian megacities in developing countries due to the subtropical climate and high-rise built context. For instance, studies in North American and European countries are mostly based on their experimental sites in low- or mid-rise built environments; whereas, studies in Japan investigating UHI in high-rise developments are mostly in non-residential sectors. Studies in Hong Kong and Singapore are based on high-rise, high-density settings but are located in tropical or quasitropical climate zones; therefore, the results must be validated in cities with subtropical climates.

Hence, answers to the following research questions, which describe the fundamental requirements for climate-responsive design, are urgently needed:

- How does urban form influence summertime UHI patterns in high-rise residential neighborhoods during the day and at night?
- What are the pedestrian wind environment and comfort conditions in high-rise residential neighborhoods during the summer months and winter months?
- How are various high-density urban geometries affecting summertime pedestrian radiant environment and thermal comfort?
- What are the environmental effects of urban greenery in modifying urban/building climates at various scales?
- How can the knowledge gained from answering the previous questions be used in the planning and design?

This book consists of eight chapters and is divided into two parts. Part One covers the theoretical and contextual background of the research. This chapter introduces the overarching topic, the nature, and the scope of the research question. Issues regarding the relationships between urban form and urban climates as well as between the thermal environment and building energy use and comfort are briefly discussed. Information on the climate and urban context of Shanghai is also outlined. Chapter 2 reviews the state-of-the-art in urban climate studies, including the concepts, theories, methods, technologies, major research findings, principles, and rules of thumb for design implications. Chapter 3 firstly briefly reviews the history and precedents of high-rise urbanism and looks into the contemporary development of high-rise urbanism in Asian megacities by comparison in economic, social, and environmental aspects of three Asian megacities, Singapore, Hong Kong, and Shanghai. The comparison indicates that a more compact high-rise urban form may better support environmental sustainable development of Shanghai, and an integrated climate-responsive planning and design workflow are to be developed. For that, better communication is needed between the disciplines of urban climatology and urban design. A framework is proposed for microclimate research and design integration for high-rise urbanism, and parametrization of urban form on microclimate at three scales is listed for the followed case studies in Part Two.

After setting the scientific background of urban climatology in Part One, the book examines the important issues in urban microclimate emerging from high-rise urban development, through a series of empirical or semiempirical case studies, setting Shanghai as the stage. We believe that solutions to urban problems can only be effectively explored in real-world cities. Part Two has five chapters. Chapter 4 investigates the effects of urban design parameters regarding urban structure, fabric, and greenery on summertime UHI intensity in ten high-rise residential quarters in the inner city of Shanghai during summer months. Datasets acquired from fixed station as well as traverse measurement allow close examinations on daytime and nocturnal UHI development and regression analysis on the correlation between design parameters and UHI intensity. Based on the resulting empirical models, design strategies are formulated for summertime UHI mitigation. Chapter 5 empirically examines the microscale effect of urban design parameters on the summertime and wintertime outdoor pedestrian-level wind environment and users' comfort in high-rise residential districts. The underlying hypothesis is that proper design strategies can be formulated for high-rise residential neighborhoods in the so-called hot-summer cold-winter (HSCW) climate zone in China, in optimizing ventilation potential in summer, while protecting users from strong winds in winter. Chapter 6 looks at the thermal radiant environment as affected by the urban morphology, greenery, and ground surface characteristics. In particular, the thermal radiant environment and its impact on people's outdoor thermal comfort are examined. The city center of Shanghai is used to illustrate how the thermal radiant environment varies with different building height/density, greenery coverage, and street layout, based on which the thermal comfort levels of different types of urban form are compared. Chapter 7 deals with urban greenery. Greenery in high-rise urban environment provides multi-scale benefits. At the local scale, it provides accessible amenity space and contributes to heat island mitigation; at the microscale, it moderates pedestrian-level thermal comfort level; and at the building scale, facade greening or so-called vertical greening system (VGS) mediates building heat gain and improves indoor comfort and energy performance. This chapter synthesizes the findings of a series of investigations on the thermal environmental performances of urban greenery at the local-, micro, and building scales.

Finally, Chap. 8 introduces a thermal atlas methodology for assessing microclimate and pedestrian outdoor comfort for complex Central Business Districts (CBD) that feature more intensive land use, diverse building form and functions, higher volume of vehicle traffic, lower greenery ratio, and of course taller buildings, compared to high-rise residential districts. Thermally comfortable outdoor environment would bring economic and social benefits to the districts so it is vital to understand and evaluate the impact of CBD development on its microclimate and pedestrian comfort. The thermal atlas methodology developed for this purpose is based on empirical modeling and urban morphology analysis on a GIS platform. The methodology is briefly introduced, and Shanghai Lujiazui CBD, the culmination of CBD development in China, is assessed by the thermal atlas as a case study. A second case in this chapter studied the Lujiazui Elevated Walkway (LEW). Elevated walkway system is becoming an indispensable way to connect land plots and buildings separated by driveways and avenues in high-rise CBD. The LEW study makes an assumption that there will be more breezes on elevated levels compared to sidewalks at the ground levels, but there can be higher exposure to solar radiation and thus higher radiant temperatures if not properly shaded. The overall effect of the change in elevation on human thermal comfort is thus unknown and needs investigation. It is to complement the thermal atlas system which is currently limited for the ground-level microclimate assessment.

References

Akbari H, Konopacki S (2005) Calculating energy-saving potentials of heat-island reduction strategies. Energy Policy 33:721–756

Bureau Shanghai Municipal Statistics (ed) (2008) Shanghai statistical yearbook. China Statistical Press, Shanghai

Che S (2004) Structural analysis and eco-planning of urban green spaces—using Shanghai as an example. The South-East University Press, Nanjing, China

Chow SD (1992) The urban climate of Shanghai. Atmos Environ 26B:9–15

Chow SD, Su J (1994) Urban Climatology (城市气侯学). Meteorology Press, Beijing

Chow SD, Zheng J, Wu L (1994) Solar radiation and surface temperature in Shanghai city and their relation to urban heat island intensity. Atmos Environ 28:2119–2127

Ding J, Zhang Z, Xi H (2002a) A study of the high temperature distribution and heat island effect in the summer of Shanghai area. Chin J. Atmos Sci 26:412–420

Ding J, Zhou H, Ye Q (2002b) Importance of City Green by investigation on evolution of heat island in Shanghai City (从上海市热岛演变看城市绿化的重要意义). Meteorol Mon 28:22–24

EPA (2002) Reducing urban heat islands: compendium of strategies. Environmental Protection Agency (EPA) (ed)

Erell E, Pearlmutter D, Williamson T (2011) Urban microclmate: designing the spaces between buildings. Earthscan, London

Fanger PO (1970) Thermal comfort: analysis and applications in environmental engineering. McGraw-Hill, New York

Gehl J (2010) Cities for people, 1st edn. Island Press

Givoni B (1998) Climate considerations in building and urban design. Van Nostrand Reinhold, New York

Glicksman L, Lin J (2006) Sustainable urban housing in China: principles and case studies for low-energy design. The Netherlands, Springer, Dordrecht

Gonçalves JCS, Umakoshi ÉM (2010) The environmental performance of tall buildings. Earthscan, London

Hang J, Li Y, Sandberg M, Buccolieri R, Di Sabatino S (2012) The influence of building height variability on pollutant dispersion and pedestrian ventilation in idealized high-rise urban areas. Build Environ 56:346–360

Japan Sustainable Building Consortium (2005) Evaluation manual for the comprehensive assessment system for building environmental efficiency, CASBEE-HI

Jones P, Storey D (2017) Density, sprawl and sustainable urban development: perspectives from the Asian and Pacific region. In: Bay JHP, Lehmann S (eds) Growing compact: urban form, density and sustainability. Earthscan, pp 56–68

Kimberly AG, Finster ME (1999) The urban heat island, photochemical smog, and Chicago: local features of the problem and solution. Northwestern University, Evanston, IL

Kubota T, Miura M, Tominaga Y, Mochida A (2008) Wind tunnel tests on the relationship between building density and pedestrian-level wind velocity: development of guidelines for realizing acceptable wind environment in residential neighborhoods. Build Environ 43:1699–1708

Li R, Wang Z, Cui L, Fu H, Zhang L, Kong L, Chen W, Chen J (2019) Air pollution characteristics in China during 2015–2016: spatiotemporal variations and key meteorological factors. Sci Total Environ 648:902–915

Li Z-Y (2005) A discussion on the sunlgiht rule for residential building in the city center. Urban Plann Forum 155:79–82 (in Chinese)

Lynch K (1984) Good city form. MIT Press

MOHURD (Ministry of Housing and Urban-Rural Development) (1994) Standard of climatic regionalization for architecture 建筑气候区划标准. GB50178-1993. China Architecture and Building Press, Beijing

MOHURD (Ministry of Housing and Urban-Rural Development) (2014) Code for fire protection design of buildings. GB 50016-2014. China Architecture and Building Press, Beijing

MOHURD (Ministry of Housing and Urban-Rural Development) (2016) Design and planning code for urban residential sectors. GB 50180-93 2016 Edition. China Architecture and Building Press, Beijing

MOHURD (Ministry of Housing and Urban-Rural Development) (2019) Evaluation standard for green building. GB 50378-2019. China Architecture and Building Press, Beijing

Ng E, Yuan C, Chen L, Ren C, Fung J (2011) Improving the wind environment in high-density cities by understanding urban morphology and surface roughness: a study in Hong Kong. Landscape Urban Plann 101:59–74

Oke T, Mills G, Christen A, Voogt J (2017) Urban Climates. Cambridge University Press

Oke TR (1987) The boundary layer climates, 2nd edn. Methuen, London and New York

Oke TR, Crowther JM, McNaughton KG, Monteith JL, Gardiner B (1989) The micrometeorology of the urban forest. Philos Trans R Soc London. Ser B Biol Sci 324:335–349

Olgyay V (1963) Design with climate: bioclimatic approach to architectural regionalism. Princeton University Press, Princeton, NJ

Rosenfeld AH, Akbari Hashem, Bretz Sarah, Fishman Beth L, Kurn Dan M, Sailor David, Taha H (1995) Mitigation of urban heat islands: materials, utility programs, updates. Energy Build 22:255–265

Santamouris M (2001) Energy and Climate in the Urban Built Environment. James & James, London, UK

SCDIC (2003) The report: green survey of Shanghai. The Shanghai City Development Research & Information Center (SCDIC), Shanghai

Schär C, Vidale P, Lüthi D (2004) The role of increasing temperature variability in European summer heatwaves. Nature 427:332–336

Schiller SD, Evans JM (2000) Urban climate and compact cities in developing countries. In: Jenks M, Burgess R (eds) Compact cities: sustainable urban form in developing countries. Spon Press, London and New York, pp 117–124

Shanghai Research Center on Aging (2018). http://www.shrca.org.cn

Steemers K, Steane MA (eds) (2004) Environmental diversity in architecture. Spon Press, New York

Stewart ID, Oke TR (2012) Local climate zones for urban temperature studies. Bull Am Meteor Soc 93:1879–1900

Szokolay SV (2008) Introduction to architectural science: the basis of sustainable design, 2nd edn. Architectural Press, U.K, Oxford

Taha H, Chang S, Akbari H (2000) Meteorological and air quality impacts of heat island mitigation measures in three U.S. cities. LBNL-44222 Lawrence Berkeley National Laboratory, Berkeley, CA

Taha H, Kalkstein L, Sheridan S, Wong E (2004) The potential of urban environmental control in alleviating heat-wave health effects in five U.S. regions: Paper J 4.3. In: 16th symposium on biometeorology and aerobiology. American Meteorological Society. Vancouver BC, August 2004

Tan J, Zheng Y, Song G, Kalkstein LS, Kalkstein AJ, Tang X (2007) Heat wave impacts on mortality in Shanghai, 1998 and 2003. Int J Biometeorol 51:193–200

Tzonis A (2006) Rethinking design methodology for sustainable social quality. In: Bay J-H, Ong B-L (eds) Tropical sustainable architecture—social and environmental dimensions. Architectural Press, Oxford, UK, pp 17–28

USGBC (2005) LEED Rating systems: LEED for New Construction, V2.2. U.S. Green Building Council

Wangpattarapong K, Maneewan S, Ketjoy N, Rakwichian W (2008) The impacts of climatic and economic factors on residential electricity consumption of Bangkok Metropolis. Energy Build 40:1419–1425

WCED (1987) Our common future. Oxford University Press, Oxford

Yeang K (1994) Bioclimatic skyscrapers (with essays by Alan Balfour and Ivor Richards). Artemis, London

Chapter 2
Urban Climates: Theories, Approaches, and Design Implications

Abstract Urban climate is characterized by a combination of physical, chemical, and biological processes occurring at different scales. From the perspective of urban living and urban design, the most relevant characteristics of urban climate include concepts such as urban heat island, urban wind, energy consumption, thermal comfort, and so on. In this chapter, the first part lays out the fundamental landscape for the conceptualization of urban climate as how to understand the climatic phenomena, characterize the physical processes, and evaluate the impact imposed by design practice. The second part focuses on a subset of urban climatology, the human biometeorology, which is most related to design practice and its implications on climate-sensitive and bioclimatic design.

2.1 Urban Climates: Understanding, Modeling, and Predicting

Both climate change and urbanization are inevitable. Climate scientists have predicted that summer temperature in high-latitude Europe will increase by at least 2 °C by 2100 if greenhouse gas emissions are at a medium level. This presents a formidable challenge for adapting cities to future extreme climatic variability such as heat waves. City dwellers have drastically modified the natural landscape in cities and created a unique kind of climate system, the urban climate. Luke Howard, a chemist and amateur meteorologist living in London back in the 1800s, was considered the first one to measure the air temperature difference between city and countryside through a scientific approach. The systematic study was well-documented in his seminal book The Climate of London, and the phenomenon observed is now called the urban heat island (UHI) effect, which is one of the most widely discussed topics about cities. Because of his pioneering work, today the annual award given by the International Association for Urban Climate (IAUC, http://www.urban-climate.org) to a scholar who has made outstanding contributions to the field of urban climatology is called the Luke Howard Award.

As Howard and his successors such as The Urban Climate by (Landsberg 1981) have shown, the climatic conditions in metropolitan areas (commonly referred to as

F. Yang and L. Chen, *High-Rise Urban Form and Microclimate*,
The Urban Book Series, https://doi.org/10.1007/978-981-15-1714-3_2

"urban") are different from that of the surrounding suburban and rural areas. Differences can be found in air temperature, humidity, amount of solar radiation received, wind speed and direction, air quality, cloud cover and amount of precipitation, etc. Due to the rapid urbanization pace globally, metropolitan areas are expanding pervasively and the boundary between "urban" and "suburban/rural" is becoming increasingly vague. Nevertheless, characterizing the effect of urbanization on the climatic variables has been a long-term pursuit for urban climatologists. Oke (1987) has set up the foundation for such reasoning through a physical approach and enlightened the major research vein in urban climatology ever since. Thanks to their contributions, today it is common wisdom that the differences of the climatic conditions found in high-density areas and that in less built-up areas can be attributed to the altering of the urban surface through the construction of artificial structures and the modification of land cover. For example, in a typical downtown environment, high-rise skyscrapers, large areas of paved so impervious ground surfaces, and lack of vegetation cover will undoubtedly change the local temperature, shadow pattern, airflow, and precipitation runoff.

Indeed, Grimmond et al. (2010) have summarized the five main categories of urban climatic characteristics that are most related to urban living. They are (1) thermal, such as the UHI effect; (2) wind, which is related to ventilation and air pollution dispersion; (3) energy, which is the energy balance and partition of the urban surface; (4) anthropogenic emission, including anthropogenic heat and CO_2 emission; and (5) water and hydrology, which is related to precipitation and groundwater balance. Understanding the cause–effect of urbanization on these key characteristics continues to be the endeavor for urban climatologist. Based on these understandings, the first two characteristics are most relevant from the perspective of urban design, as they can be largely manipulated by design actions. So these two factors will be introduced in this chapter. But before embarking on the discussion, let us first take a look at the different spatial scales that an urban climate works on. Understanding the working mechanism for an urban climate at different scales is a key to characterize the physical phenomena and processes involved and transfer such knowledge to actionable measures in design terms.

2.1.1 *Tackling with Different Scales*

The climate system is a result of the complex interactions between the atmosphere and the earth's surface, mostly because the atmosphere's constituting unit, the air particle's velocity, and viscosity features are engaged in two different kinds of physical processes. As a result, a "boundary" emerges where the atmosphere meets the surface. In general, atmospheric motions range in size from about a millimeter to tens of thousands of kilometers horizontally and the entire depth of the atmosphere vertically. Three boundary layers are normally used to describe such complex circulations at different spatial scales with respect to cities: the planetary boundary layer (PBL), the urban boundary layer (UBL), and the urban canopy layer (UCL). And accordingly, three different spatial scales are recognized to distinguish atmospheric processes in urban areas and the atmospheric layers: the mesoscale, the local scale, and the microscale.

The lowest level (about 10%) of the troposphere is known as the PBL and normally considered to be around 10 km in size in the vertical direction. This is the layer that most meteorological phenomena related to mankind take place and the weather forecast works on. The corresponding mesoscale urban climate is normally in the order of hundreds of kilometers horizontally consisting of a metropolitan area or urban agglomerations, so the atmospheric motion is essentially a part of the regional climate. The bottom of the PBL above cities is recognized as the UBL and is normally 1–2 km in size in the vertical direction. The atmospheric motions in this layer are significantly influenced by the high surface roughness of the urban structure and the thermal turbulence induced by the thermal load of buildings and anthropogenic heat emissions. Within the UBL, the urban climate is considered to be the local scale, ranging from one to several kilometers horizontally, i.e., consisting of a district, a neighborhood, a set of blocks, an array of houses, etc. The bottom of the UBL is the UCL where the impact of the local urban settings on the climatic conditions peaks. UCL extends from the ground level approximately to the top of buildings, which makes its vertical dimension vary a lot for different cities, ranging from several meters to several hundred meters. Within the UCL, the local-scale urban climate is strongly modified by the thermal properties of buildings and ground surfaces and the airflows arising from the buildings' geometry and streets' fabrics (Oke 1987). The microscale of the urban climate dominates in the urban canyons and configurations created by individual buildings or building clusters, trees, and spaces, etc. The climatic conditions in such microscale could be quite different from the local-scale urban climate, such as wind velocity and direction. Typical microscales range from one to several hundred meters.

2.1.2 Heat and Wind

Heat and wind are the two aspects of the urban climate that cities have the most profound impact on. In fact, the heat and wind conditions at the microscale can be greatly modified by urban configurations in cities. Therefore, they are of particular interest for architects and town planners.

- UHI and Beyond

The UHI effect has been the most widely documented and examined topic in urban climatology and continues to draw attention from research practitioner communities (Arnfield 2003). While the traditional notion for UHI is that the nocturnal air temperature in a city center can be up to several degrees higher than that in its rural surroundings, this concept has also been pushed forward due to the rapid urbanization pace and high-density living trend (see Chap. 3 of this book) and also our advancing understanding of the various causality of the UHI formation. Today, it is commonly recognized that there are two types of heat island in cities, the conventional air temperature heat island and also the surface heat island (SUHI) where the surface temperature in urban settings is significantly higher than that in suburban and rural areas, meanwhile showing a much diversified spatial pattern compared with the air temperature variations. Although not as thoroughly studied as UHI, SUHI has become of increasing interest in urban climatology with the progress in remote sensing techniques and availability of geospatial database. It is to be noted that with the rapid expansion of urban boundary, the compelling focus of urban heat island study has arguably shifted from urban–rural comparison in the mesoscale to intra-urban differentials in the microscale. For the sake of consistency, this chapter uses the term of UHI in its most generic sense, the temperature difference measured in developed areas and less built-up areas in cities (Fig. 2.1).

The formation of the UHI is determined by the energy balance of the urban surface, which is defined by the following equation suggested by Oke (1988) assuming an imaginary volume of an urban canyon in the UCL (Eq. 2.1):

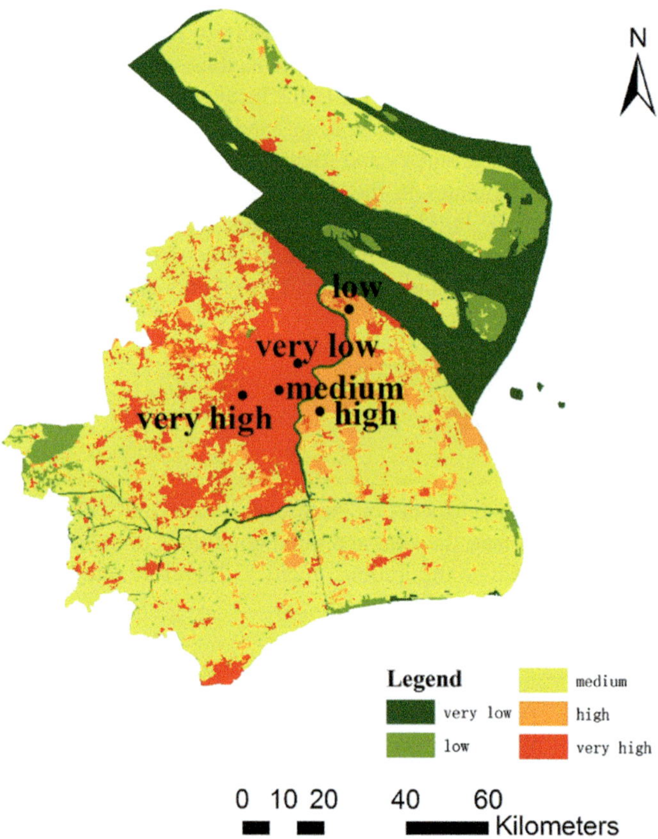

Fig. 2.1 SUHI over Shanghai

$$Q^* + Q_F = Q_H + Q_E + Q_S + \Delta Q_A \tag{2.1}$$

where Q^* is the net radiation flux received by the urban surface, including both shortwave and longwave radiations, and is essentially the energy available to "heat up" the local air if other climatic conditions allow. Q^* can be further decomposed as (Eq. 2.2):

$$Q^* = K^* + L^* = K\downarrow - K\uparrow + L\downarrow - L\uparrow \tag{2.2}$$

where K and L stand for shortwave radiation flux (known as solar radiation) and longwave radiation flux (known as terrestrial radiation), respectively. Therefore, K^* is the net shortwave radiation flux, $K\downarrow$ is the downward shortwave radiation flux received by the surface, $K\uparrow$ is the shortwave radiation flux emitted from the surface to the sky, and similar notations apply for L.

Q_H is the turbulent sensible heat flux, meaning this is the part of energy that can be observed and measured, such as the energy heating up the air or the ground. Q_E is the turbulent latent heat flux, meaning the part of energy absorbed or released during constant-temperature processes, so it cannot be directly observed, i.e., phase transitions such as evaporation. Q_S is the storage heat flux representing all sensible heat budget in the volume, including the heat storage in air, water, vegetation, building surface, soil, etc. ΔQ_A is the net advection heat flux through the sides of the volume and can often be neglected for microscale cases. Q_F is a very important term which is unique for urban surfaces but not present in other landscapes. It is the anthropogenic heat flux induced by energy consumption, such as fuels (traffic, industry, etc.), electricity (air conditioning, lighting, etc.).

It is easily found from the energy balance equation that the UHI is essentially the result of differences in the energy balances for urban and less built-up areas. In daytime, urban ground and wall surfaces with lower albedo, such as asphalt and tar absorb more shortwave radiation. With less vegetation cover, the surface has less moisture, so less energy is dedicated to latent heat and more is available for heating up the air and the ground. This process is further enhanced by the abundant anthropogenic heat input, and as a result, a large partition of the shortwave radiation is stored in the urban canyons. At nighttime when the sun is absent so the ground is warmer than the air, the urban structures greatly delay the cooling process of the ground by "trapping" the energy in the form of longwave radiation between buildings, and the air volume within in the urban canyon is effectively heated; therefore, higher temperature can be observed. A quite related concept to characterize such processes is the sky view factor, often denoted by Ψ_{sky}, which is the proportion of the sky hemisphere visible from ground level. It is largely determined by building height, volume, and layout. Sky view factor has an important role in the energy balance of urban surfaces, and a general rule is that larger sky view, i.e., less obstructed sky, indicates larger escape of longwave radiation from the ground to the sky and hence faster cooling process, i.e., lower temperatures. Apart from the geometric aspects of the urban structures, the wall materials and ground surface characteristics will also greatly alter the different partitions of the energy balance in urban areas and eventually

Fig. 2.2 Fish-eye lens photographs taken at ground level showing different levels of sky views. Left: $\Psi_{sky} \sim 0.95$, right: $\Psi_{sky} \sim 0.35$

the microclimatic condition, which is related to the thermal radiant environment and will be further discussed in Chap. 6 of this book. To sum up, urban geometry, land cover characteristics, and also human activity can significantly affect the microscopic atmospheric conditions in UCL. This suggests that interventions such as architecture and landscape design can be utilized as countermeasures to mitigate UHI (Fig. 2.2).

- Turbulence and Wind in Cities

Turbulence is the defining characteristic of the PBL. Turbulence in PBL occurs because of the shear in the mean wind and is further complexed by the temperature stratification, i.e., being either amplified or restrained. In fact, alteration to the wind field both in the vertical and horizontal dimensions is one of the most significant urban effects on the climatic condition in the UBL and the UCL. This is reflected in the extremely localized wind speed and direction in the lower few hundred meters of the atmosphere. The difference occurs as a result of the increased frictional drag on air flowing over built-up urban terrain, which is rougher than rural areas. Roughness plays a central role in affecting the local wind flow, and it refers to all sorts of obstructions, including building bulks, street networks, tree canopies, vehicles, and even signage and banner boards if the microscopic scale is examined. Consequently, the wind speed is largely reduced in the low level of UBL due to the increasing roughness. Another noticeable difference between urban and rural wind flow is the convergence of wind in the lower layer of UBL; i.e., air tends to flow into built-up areas from all directions. This effect can be primarily attributed to the horizontal thermal gradients of the urban heat island intensity, and as a result, the low-level wind field across cities becomes extremely complicated.

In urban wind engineering and related fields, vertical profiles of wind speed and turbulence are of great importance. Understanding how the wind speed near the ground is affected quantitatively by artificial structures can be very helpful for architects and planners aiming at introducing ventilation to high-density cities especially for the designing and planning of high buildings and towers. By bringing more wind flow into well built-up city centers, the urban heat island can be largely mitigated, and air pollutants that are emitted or secondarily formed near the ground can also be effectively dispersed.

For neutrally stratified UBLs, i.e., with purely mechanical turbulence, the wind velocity u at height z is commonly given by the classical and semi-empirical logarithmic law derived using data of field observations of wind or turbulence (Eq. 2.3):

$$\frac{\bar{u}(z)}{u^*} = \frac{1}{\kappa}\ln\left(\frac{z - z_d}{z_0}\right) \tag{2.3}$$

where $\bar{u}(z)$ is the time-averaged wind velocity in z height above ground; u^* is the friction velocity (or shear velocity) at the ground surface and is proportional to the square root of the local Reynolds stress; κ is von Karman's constant and usually designated to 0.4 (Counihan 1975).

The two major aerodynamic parameters, the aerodynamic roughness length z_0 and the displacement height z_d, are of great importance in the study of the wind profile. z_0, as the name implies, has the dimension of m and attributes the "roughness gradient" of the surface. It is purely dependent on the terrain type and is primarily related to the ratio of the frontal area of the obstacles to the lot area of the obstacles. The estimation and classification proposed by Davenport et al. (2000) for different types of terrains have been broadly used. The table gives a summary of the roughness lengths of different landscapes and their classifications (Table 2.1).

The displacement height z_d identifies the level of the aerodynamic surface where $\bar{u}(z)$ obtained from the logarithmic wind profile goes to zero if the log-law was

Table 2.1 Davenport classification of terrain roughness

Classification	Landscape	Roughness length z_0
Sea	Sea, lake, desert	0.0002
Smooth	Beach, ice, open country	0.005
Open	Low vegetation	0.03
Roughly open	Cultivated land	0.10
Rough	High crops	0.25
Very rough	Forest	0.5
Closed	Density built-up area	1.0
Chaotic	City center	2 and above

Abridged from (Davenport et al. 2000)

maintained from the outer flow all the way down to the surface. Below, this level is the lower part of the roughness sub-layer (RSL) where the log-law is no longer applicable. It is important to include z_d in the study of urban wind profile, as it essentially sets a new "ground surface" and therefore a new base for the log-law to recognize the impact of the physical bulk of the urban canopy (Oke 2006a). In cities, this base is normally considered equal to 0.5–0.8 of the average height of the surface roughness elements, i.e., average building height depending on the building and tree density (Grimmond and Oke 1999). By definition, z_d is mainly a function of the ratio of the plan area of the obstacles to the lot area.

A widely used method to estimate the wind velocity within the RSL is to use a power-law formulation (Eq. 2.4):

$$\frac{u(z)}{u_r} = \left(\frac{z}{z_r}\right)^{\alpha} \tag{2.4}$$

where $u(z)$ is the average wind velocity at height z, u_r is the wind velocity at reference height z_r, and α is the wind shear exponent and is related to the surface roughness z_0. In real cities, the size and shape of roughness elements are normally quite irregular and variable. This makes the estimation of z_0 particularly challenging. A commonly used approach is to consider the front area index (FAI, denoted as λF) of all buildings in a site. As the name implies, λF is the total area of the building wall façades below a certain height projected to the plane perpendicular to the approaching wind direction. For a specific area, the spatially averaged sum of λF is called the frontal area density (FAD) and is equal to z_0 when the height of the UCL is considered.

With the approaches and technics described above, wind speed at a height within several hundred meters above ground can be estimated in a general sense. However, down into the street canyon level, the wind flow can be drastically different from the ambient wind profile and turbulence is the dominating factor. A general rule of thumb is to use the aspect ratio (AR) to distinguish different patterns of wind. AR for street canyons is defined as the ratio of height (H) to separation distance or simply street width (W). Therefore, the wind flow inside street canyons can be classified into different flow regimes depending on the H/W range. The threshold values estimated by (Oke 1988) have been commonly adopted, i.e., isolated roughness flow with $H/W < 0.3$, wake interference flow with $0.3 < H/W < 0.6$, and skimming flow with $H/W > 0.6$. Although not an ultimately accurate capture, this estimation provides practitioners especially those without domain knowledge in meteorology or climatology an easy way to interpret how street and building configurations determine the wind flow pattern in street canyons, and further impacts, such as the dispersion behavior of pollutant, can be examined.

2.1.3 *Understanding Phenomena and Processes from a Design Perspective*

As has been discussed above, the climate in cities is characterized by a series of complex phenomena imposed by a combination of processes occurring at different spatial and temporal scales. Understanding of such phenomena and processes based on observation of relevant climatic variables and also analytic approaches will greatly enhance our capability to address climate issues in urban design practice and incorporate respective climatic variables in the decision-making process. Indeed, Oke (2006b) described the evolution of urban climate studies using eight modes of investigation or practice: conceptualization, theorization, field observation, modeling, model evaluation, application in urban design and planning, impact assessment and policy development and modification. While the later three modes concern implementation and application in the real case, the former five modes together layout the scientific background for investigating urban climate issues by constituting a complete logistic loop for scientific reasoning and experiment methodology design and help to foster our understanding of these issues in different scales.

The most fundamental way to understand the phenomena and processes involved in urban climate, as along the history of the wisdom of mankind, is through field observation. It provides us with fundamental dataset upon which scientific questions can be raised and analytical inquiries can be made. Qualified field observation in cities should have the following attributes: fixed location based, making the results geographically and climatically related and eligible for longitudinal comparison studies; long term based, meaning the dataset achieved can be temporally incremental and therefore can reflect temporal evolution and the impact of climate change; and sustainable, i.e., the methodologies and protocols employed can be constantly updated and extended, allowing continuous contribution to the research and practice community.

The Basel UrBan Boundary Layer Experiment (BUBBLE) is one of the most representative examples from this perspective. BUBBLE was initially a one-year experiment project (2001–2002) jointly operated by the University of Basel, ETH Zürich, University of British Columbia, Western University, and NUS (Singapore) to study the microclimate such as airflow, energy exchange, and air quality in cities (Rotach et al. 2005). The approach is to investigate in detail the structure of the urban boundary layer in the city of Basel, Switzerland. A network of multiple field observation stations, i.e., a total number of 7 tower sites were installed to examine the turbulence (i.e., wind flow), turbulent fluxes (e.g., heat flux), and energy exchange (e.g., radiation budget) features over different urban surface types, namely urban, sub-urban, and rural. All sites were operated simultaneously, and the measurements were conducted at multiple levels for a long term which provides detailed profiles of the urban boundary layer both in the horizontal and vertical dimensions. The dataset obtained from BUBBLE was made publically available to the urban climate research community with no restrictions and has been widely used to validate numerical models such as urban radiation models, urban energy balance models, and urban canopy parameterizations. An experimental established nearly 20 years ago, the framework of BUBBLE has also been substantially extended, techniques such as Lidar wind profiler and tethered balloon were performed for detailed investigations in street canyons and the project is still active today. Long-term field observations of atmospheric conditions and processes in cities of this kind have been widely installed around the world and provide the fundamentals for advances in the understanding of urban impact on climate. The progress in the development and practice of such projects has been summarized in (Grimmond 2006).

2.1.4 *Models and Predicting Tools*

While field observations provide us with necessary dataset, models and analytical approaches alike can push the frontier of the understanding and reasoning of urban climates. From a design perspective, understanding the cause–effect of design actions on related climatic variables is virtually of more concern than figuring out the underlying mechanism of atmospheric processes. Givoni et al. (2003) addressed the need for "predicting tools" in the research for how changes in design details influence outdoor thermal comfort which can be further extended to climate-sensitive design in general. As they put it, "In order to evaluate the importance of modifying the outdoor climate in a particular direction by specific design details it would be helpful if the designer would have some means for 'predicting' the effect of a particular change in a climatic element on the comfort of persons staying outdoor." Models contribute significantly to the understanding of urban impact on climate. Customized models incorporated in a decision-making loop can serve as a predicting tool to inform practitioners how the city will respond to such design scenarios.

One of the most straightforward uses of models is to estimate, and for complex climatic phenomena such as the UHI, scale models are commonly used to simplify, describe, and characterize the processes and relationships involved. A scale model in the loosest sense is a physical representation of the object of interest that maintains accurate relationships between all important parts of the model and simplifies the investigation of the original object by looking at the reconstructed variables and relationships in the model, normally in reduced spatial scales. Oke's scale model (1981) is an example that cannot be circumvented when discussing fundamental models in urban climatology. He characterized the formation of nocturnal UHI by building a simplified physical scheme based on a transient cooling model and incorporated the effect of urban geometry as described by the spatial parameter of SVF for estimating the longwave radiative heat transfer. Scale model has been widely used to determine the contribution of heat transfer processes to the UHI, the characteristic temporal and spatial variation of surface temperature, and also the impact of urban obstacles on the flow pattern. Complex urban geometry and terrain and landscape properties can be simplified using spatial indicators such as SVF or empirically determined parameterization and incorporated in such models. They are especially informative to systematically investigate the relationships between surface structures and physical processes for complex urban settings. Kanda (2006) has provided a seminal and comprehensive review of the scale modeling methodology in urban climate study.

When the physical process becomes more complicated, such as flux turbulence and wind flow in cities, scale models will not provide enough details for a comprehensive understanding of the process. In cases, numerical modeling is a more suitable approach. Generally speaking, numerical modeling uses mathematical equations to describe the physical process, and mathematical solutions to the equations are achieved by numerical time-stepping procedure to obtain the models' behavior over time, i.e., computational simulation. The main advantage of the numerical modeling approaches compared with traditional observation and simplified scale model is that they allow comparative analyses based on different scenarios, which is essential in design practice. In addition, while observations are usually based on measurement stations and flux towers installed at a limited number of locations in space, numerical models and simulations can be performed in an entire computational domain. Indeed with the rapid increase of computer power and progress in geospatial techniques, numerical simulations have been increasingly carried out for real-case scenarios. They help to predict complex phenomena such as forecasting urban weather, air quality, precipitation, and potential extreme events and hazards.

The computational fluid dynamics (CFD) model is one of the most representative examples of how numerical models can be applied to the investigation of complex physical processes and enhances our understanding. In the past decades, CFD has been studied intensively as a tool for evaluating the turbulence flow in cities due to the increasing necessity for simulations incorporating higher spatial and modeling details. As Souch and Grimmond (2006) have put it, "The development and use of CFD is a very active area of inquiry. The models are becoming more sophisticated in terms of numerical methods, mesh structures and turbulence modeling approaches." They have been widely used to simulate urban-type flows at microscales; i.e., with CFD, it is possible to resolve the flow field at very fine spatial scales (e.g., building or even human scale). There are three main numerical approaches for solving the mathematical equations—Reynolds-averaged Navier–Stokes (RANS), large eddy simulations (LES), and direct numerical simulations (DNS), with increasing simulation accuracy but also larger demand for computational resources at the mean time. In all approaches, CFD discretizes the spatial domain into a grid or mesh of points or elements and obtains the numerical solution by approaching in discrete time steps until a satisfactory accuracy is reached. CFD simulations require high-resolution representation of the urban geometry, the knowledge of boundary conditions for all relevant flow variables, and above all sufficient computational resources. Due to CFD's large demand of computational resources, most of the studies have been based on simple and regular geometric elements such as arrays of building cubes and rectangular street canyons. Spatial indicators and morphological parameters such as H/W and λF can be used to simplify the calculation (Ng et al. 2011). The impact of the layout of buildings such as spacing and alignment and height differences of buildings on the flow pattern has been more widely discussed. On the other hand, the effects of building shapes have not drawn as much attention. Interested readers are referred to (Blocken 2015) for an up-to-date comprehensive tutorial review of CFD.

It is to be noted that the predicting tool normally consists of a combination of simple models and more sophisticated numerical models (or simulations) at different scales. Energy budget models have been traditionally used for characterizing the radiation fluxes and heat storage in street canyons (e.g., Grimmond and Oke 1999). Buildings or blocks have been the common scale of investigation. Meanwhile, computer simulation tools have been developed using geo-dataset to look at the neighborhood to city scale with various morphological features (e.g., Lindberg et al. 2008). There have also been estimating models at the global scale (Allen et al. 2011). On the other hand, how radiation fluxes affect air temperature in street canyons is a dynamic and complex process, and recent advancement in CFD makes it possible to carry out numerical simulation of the involved turbulences and flows in finer scales (Takahashi et al. 2004). The combination of the finer scale understanding from numerical simulation and the broader scale pattern from morphological characteristics is expected to bring new insights into the formation and development of urban climatic processes, especially in complex urban settings.

2.2 Urban Human Biometeorology and Design Applications

The climatic impacts of design scenarios on human beings cannot be fully understood by studying any single meteorological parameter, such as air temperature, in isolation. A link between the local climatic characteristics and human well-being is essential in providing support for the decision making in urban design and (re)development. For the past two decades, there has been a growing research interest to build this link from the urban human biometeorology perspective. Human biometeorology is the study of the influence of the atmospheric environment, such as weather, climate, and air pollutants on human organism. Its basic landscape was developed in the early 1900s, and the well-established framework has been extensively documented by the literature of the International Society of Biometeorology (Sargent and Tromp 1966). But it is only until the 1990s have the human-biometeorological studies been put into the outdoor urban context and connected with outdoor space use, human activities, and space design concern (Fukuoka 1997; Höppe 1997; Jauregui 1997). Various climatic elements are considered, including thermal, actinic, and air hygienic characteristics of the urban climate (Höppe 1997). Their effects on the health and well-being of urban populations are assessed through a physiologically significant manner (Mayer 1993). With respect to directing decision making in urban design, the assessment method of the modified human biometeorology has been formalized in practice guidelines, such as the German standard VDI guideline methods for the human-biometeorological assessment of climate and air hygiene for urban and regional planning (VDI 3787 1996). Since the urban microclimate is site-specific and can be most controlled by design choices, a great variety of design approaches have been taken to mitigate the

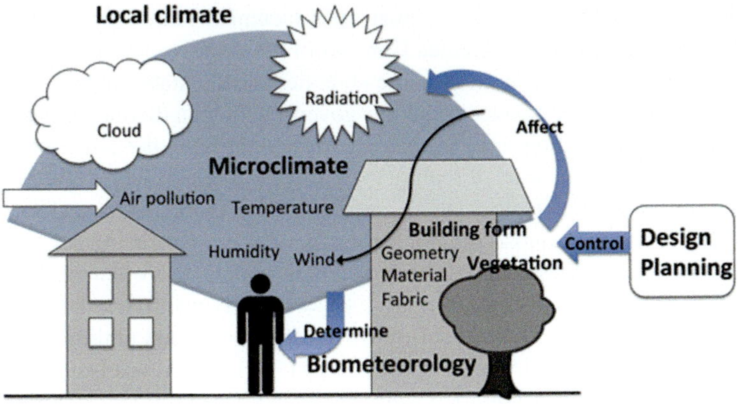

Fig. 2.3 Sketch of the elements of urban human biometeorology and how it is related to urban design and planning

impact of climate change on urban living quality and enhance city residents' adaptation to climatic variability based on human-biometeorological understandings (Chen and Wong 2006; Ali-Toudert and Mayer 2007; Gaitani et al. 2007; Holst and Mayer 2011). Figure 2.3 shows a sketch of the elements of urban human biometeorology and how it is related to various elements in urban space design. Although there have been general reviews on the development of human biometeorology as a discipline (Sargent 1965; Landsberg 1972; Höppe 1997; McGregor 2011), discussions have been sparse on how it is related with design practices compared with its rapidly growing literature. Furthermore, the justifications of the human-biometeorological research are commonly neglected in the dialog.

2.2.1 Embracing the Human Dimension: Thermal Comfort

Linking climatic characteristics with human sensation is an important task in building climate-responsive cities. By understanding which climatic variables have the most profound impact on human, design strategies can be worked out accordingly. Assessment method, in such a context, is the key. Urban human biometeorology built its theoretical foundation upon the human thermal regulatory model developed in the 1970s (Fanger 1972). A set of climatic variables, such as air temperature, radiation, humidity, air pressure, and wind speed, have been identified as having the most important effects on human sensation, especially human thermal sensation (Givoni 1976). On the other hand, these effects cannot be fully understood by examining any single climatic parameter in isolation. For example, earlier studies in Hong Kong found that a decrease in wind speed of between 1.0 and 0.3 m/s in the downtown open space can lead to a 1.9 °C air temperature increase in a thermal significant manner (Cheng and Ng 2006). Therefore, a number of biometeorological indices have been developed to assess the overall impact of various climatic parameters on human thermal sensation (Task Committee on Outdoor Human Comfort of the Aerodynamics 2004).

A major group of such indices is the so-called steady-state models, which are based on solving energy balance equations governing human thermoregulation. The indices are normally represented in scale metrics quantifying the assessment of the thermal quality. The first kind of scaling system is level based, such as the predicted mean vote (PMV) index (Fanger 1972), and the index of thermal stress (ITS) (Givoni 1976). PMV is one of the most commonly used thermal assessment indices. It predicts the mean thermal response of a large population of people. It is often measured on a seven-point scale (+3 = hot, +2 = warm, +1 = slightly warm, 0 = neutral, −1 = slightly cool, −2 = cool, −3 = cold). PMV has been included in the International Organization for Standardization ISO standard (ISO 1994). It has been commonly adopted in outdoor thermal comfort studies in which large groups of people are being surveyed (Nikolopoulou et al. 2001; Thorsson et al. 2004; Cheng et al. 2012).

The second kind of scaling is temperature-based indices, which are measured in degree Celsius (°C) and can be easily interpreted by architects and planners without sufficient knowledge about meteorology. Notable examples include the physiological equivalent temperature (PET) (Mayer and Höppe 1987) and the OUT_SET* (Pickup and de Dear 1999). PET is defined as the air temperature at which, in a typical indoor setting, the human energy budget is maintained by the skin temperature, core temperature, and sweat rate equal to those under the conditions to be assessed (Höppe 1999). PET is particularly suitable for outdoor thermal comfort analysis in that it translates the evaluation of a complex outdoor climatic environment to a simple indoor scenario on a physiologically equivalent basis that can be easily understood and interpreted. PET has been widely applied in areas with various climatic conditions (Matzarakis et al. 1999; Ali-Toudert and Mayer 2006; Lin 2009; Cheng et al. 2012).

Recently, there has been growing research interest in understanding people's dynamic thermoregulatory progress in movement and activity. This is based on the realization of people's dynamic feature and adaptive capability in outdoor environment (Höppe 2002). Several models and assessing frameworks have been proposed and experimentally applied, such as the Pierce two-node model (Gagge et al. 1986) and the Instationary Munich Energy Balance Model (IMEM) (Höppe 1984). The norm is to use a detailed set of thermoregulatory indicators, such as core temperature, skin temperature, and sweating rate and skin blood flow to describe a person's physiological condition. Although these models can provide detailed investigations of the dynamic course of human thermal adaptation, they normally involve extensive monitoring of human subjects if applied in real-world cases and require domain knowledge in biometeorology and physiology to interpret the results. Given human being's highly diverse characteristics and unpredicted movement in reality, these dynamic models would face great barriers before they can be used as decision support tools in outdoor space design. In contrast, the steady-state biometeorological indices can be easily calculated in site-specific thermal assessment, and being effective, efficient, and representable, they have been commonly applied for design-support purposes.

- Mean Radiant Temperature

The mean radiant temperature (T_{mrt}) is the most important biometeorological parameter that characterizes the effect of thermal radiant environment on human thermal comfort (Fanger 1972). It is defined as "the uniform temperature of an imaginary enclosure in which the radiant heat transfer from the human body equals the radiant heat transfer in the actual non-uniform enclosure" (ASHRAE 2001). By definition, T_{mrt} parameterizes the holistic effect of the complex radiation fluxes of the thermal radiant environment into a single temperature dimension index (°C). In a complex urban environment, the radiation fluxes vary considerably in outdoor spaces because of the shading generated by buildings and vegetation and also because of different surface materials. Therefore, T_{mrt} can exhibit significant spatial variation even within a short distance. For example, Mayer and Höppe (1987) showed through in situ meteorological measurement that the T_{mrt} difference between a sunlit street canyon and a nearby shaded street canyon can be up to 30 °C in the early afternoon, whereas the difference of the air temperature (T_a) is less than 3 °C. This suggests that T_{mrt} can be a more suitable index for characterizing the intra-urban differences of thermal comfort conditions as compared to traditional meteorological indices such as Ta, especially in complex urban environment. Therefore, T_{mrt} has been widely adopted across the world in urban human-biometeorological studies to parameterize the level of thermal comfort and heat stress in hot summer (Koch 1962; Höppe 1992; Masmoudi and Mazouz 2004; Thorsson et al. 2004; Emmanuel et al. 2007; Kántor and Unger 2011). Thorsson et al. (2014) also found that T_{mrt} can effectively predict the risk of mortality of senior citizens induced by heat stress.

In urban outdoor spaces, T_{mrt} is primary determined by building geometry, street layout, albedo of facade and ground, and vegetation cover. These components can be effectively altered by design interventions. Therefore, understanding how the spatial variation of T_{mrt} is affected by different urban settings is important and can provide valuable implications for climate-responsive urban design and planning. An increasing number of studies have hence addressed this concern across different scales, including microscale such as street canyons (Ali-Toudert and Mayer 2007; Lee et al. 2013, 2016) and urban open spaces (Chatzidimitriou and Yannas 2015), neighborhood scale such as residential communities (Lai et al. 2014), and district scale with various building shapes and street orientations (Radhi et al. 2013; Ketterer and Matzarakis 2014; Lindberg et al. 2016; Yang and Chen 2016; Yang and Lin 2016). It is commonly found that building geometry and vegetation plays the most significant role in affecting T_{mrt} and thus the overall thermal comfort (Ali-Toudert and Mayer 2007; Lindberg and Grimmond 2011; Lau et al. 2015; Lee et al. 2016); in contrast, the contribution of surface albedo is rather minor (Erell et al. 2014; Lindberg et al. 2016a). So far, most of the studies were carried out in urban environment with low or moderate density, such as mid-size European cities (Lau et al. 2015; Lee et al. 2016; Lindberg et al. 2016b). In the high-density urban environment of a metropolitan such as Shanghai, the spatial variation of T_{mrt} is further complexed by the combination of deep street canyons, shadow casted by high-rise buildings, near-wall sunlit spaces, etc. Lau et al. (2016) took the downtown center of Hong

Kong as an example and addressed the issue in high-density urban environment. On the other hand, the discussion was qualitative rather than analytical. No quantitative comparisons were given for the spatial variation of T_{mrt} of the selected "open area" and "denser area" which are separated by a road, and no vegetation was included in the estimation of T_{mrt} even though there were urban parks and green spaces, which are also essential urban elements. Furthermore, the use of summer average weather data may be less revealing since Thorsson et al. (2014) have shown that daily maximum T_{mrt} can better reflect the level of heat stress as indicated by mortality. Chapter 6 of this book presents a real case study using T_{mrt} to characterize the urban thermal environment in a densely built urban center (Shanghai).

2.2.2 Design with Climate: Passive and Bioclimatic Design

Buildings constitute the largest section of resource consumption and heat emission in cities. As the name implies, passive building is in contrast with the active or mechanical building, which depends on artificial input as operational power. In comparison, in a passive design, the building and the spaces within it are designed to benefit from natural light, ventilation, and temperatures. The term "passive solar architecture" has long been used to describe buildings that harness sun energy for the provision of comfort condition—typically to provide warmth (Brown and DeKay 2001). In Europe, theorists and researchers have gone a step further to use the term "passive architecture" to describe buildings that make use of the envelope for the provision of a "selective environment" under different climatic conditions (Hawkes et al. 2002). Passive buildings can greatly enhance a city's adaptive capability to climate change by reducing energy use and environmental impacts such as greenhouse gas emissions and therefore increase a city's resilience in extreme weather condition. The harmony with natural environment is physiologically more appealing to the human body and therefore can contribute to human comfort and well-being.

As such, traditional and vernacular buildings, which respond to the natural environment, are regarded as the provision of wisdom of what climate-responsive buildings should be. Givoni, an early proponent of passive and bioclimatic design and one of the founders of the Association of Passive Low Energy Architecture, has contributed the technical understanding of the concept in his seminal book Man, Climate and Architecture, in which he summarizes, "[passive design] involves architectural design and choice of materials aiming at providing comfort while minimizing the demand for energy use…Special features of design and of structural materials affect the response of a building to exposure to climatic elements: the quantity of solar radiation absorbed in and penetrating the building, the air and surface temperatures, the air velocity and the vapor pressure" (Givoni 1976).

Olgyay's bioclimatic understanding on passive design has long been regarded as a theoretical model in building design (Olgyay 1963). Being passive allows building to harness solar energy and the natural resources in the surrounding environment, thus to reduce the use of space heating and air conditioning while maintaining comfort of building occupants. The better the passive building performance, the more natural resources from surrounding environment could be used to reduce energy consumption. It depicts that mechanical heating or cooling measures "flatten" the temperature curve of buildings. In contrast, passive buildings or climatically designed buildings offer a range of potential temperatures that go along with the environmental conditions, saving a lot of operational energy and at the same time giving occupants opportunities to adjust.

The potential of a nature/building relationship has been explored in (McHarg 1992). A passive building responds to local climate and site conditions. Scholars and practitioners have always emphasized the importance of regional variations in passive design. Typically, there are three types of climatically dependent environmental parameters that could be utilized with passive design and have been widely considered in research and practice, including sun, daylight, and wind (Givoni 1994; Brown and DeKay 2001; Szokolay 2004; Kwok and Grondzik 2007). For example, a passive building provides shading devices to protect from radiation, and by shaping them according to the changing seasonal sun path, both summer and winter heat gain can be achieved; a passive building also maximizes the use of natural light to light the room, both for energy efficiency and for the health and comfort of occupants; and in a passive building, desirable air movements are utilized for cooling in hot periods and as a relief from vapor pressure during times of high humidity. Passive design strategies under different climatic conditions is summarized (Jones 1998).

Human biometeorology provides understandings to improve human comfort and well-being by optimizing the urban form. The general term of urban form here refers to the broad urban landscape consisting of building, street, space, greenery, etc., which largely determines the local microclimatic condition. Bioclimatic design responds to the local climate and site conditions to improve the urban thermal environment, wind flow field, and ultimately human thermal comfort (Swaid et al. 1993; Jones 1998; Nikolopoulou 2004; Gaitani et al. 2007). In outdoor spaces, human is directly exposed to the ambient environment, solar radiation, daylight, and wind and therefore can most benefit from design choices that optimize building form and street furniture to improve the local microclimate for better thermal comfort. Bioclimatic design principles have been widely applied in hot regions to promote human thermal comfort sensation and save energy consumption for mechanical cooling. Air temperature and solar radiation are under all conditions the two most important factors in regulating human biometeorology. In hot and arid climates, much attention has been addressed to the protection from excessive sun exposure. In hot and humid regions, improving air movement is a primary concern in order to make city residents feel cooler (Masmoudi and Mazouz 2004; Ali-Toudert and Mayer 2007). Bioclimatic design can also utilize desirable air movements for cooling in hot periods and as a relief from vapor pressure during times of high humidity. For example, Ahmed (2003) found that with wind speed above 2 m/s, the level of humidity for comfort was found to extend up to 95%. Similar results are also shown in (Stathopoulos 2006). Other studies have been carried out in regions with moderate climates (e.g., Spagnolo and De Dear 2003; Bourbia and Boucheriba 2010). Biometeorological indices such as PMV (Vanos et al. 2010) and OUT-SET* (Spagnolo and De Dear

2003) were employed. Sometimes, more detailed human-biometeorological parameters were used instead of a comprehensive index (Gaitani et al. 2007). Tseliou et al. (2009) compared three biometeorological indices, namely PET, the temperature-humidity index (THI), and the wind chill index, to the surveyed actual sensation vote (ASV). The results revealed that the thermal comfort classes as calculated by various biometeorological indices have a strong correlation with the mean temperature of the study area. The authors attributed this correlation to the acclimatization and adaptation, indicating that human thermal comfort sensation depends largely on the how they have adapted to the ambient thermal environment.

In his eight-mode depiction of the communication between urban climate research and urban design, Oke (2006b) opined that whereas the first four modes (conceptualization, theorization, field observation, modeling) had progressed significantly in recent years, the latter four modes (model evaluation, application in urban design and planning, impact assessment and policy development and modification) remained undeveloped in terms of transferring domain knowledge into design actions, and today, it is still not common practice to take climatic considerations in design and development projects. This is especially true in the early design process where many of the major design decisions take place in terms of neighborhood layout and building geometry. Designers and policy makers need simple instructions and guidelines to support early-phase design decisions that affect the microclimate (Eliasson 2000). For example, if wind flow evaluation is to be carried out in a design project, predicting tools such as CFD are either too time-consuming or too technical (and most of the time both) to be suitable for early-phase evaluation performed by urban designers. Design recommendations and guidelines which can be easily interpreted without expertise in climatology would quickly provide users with relevant outputs without substantially impacting their existing design process (Brown and DeKay 2001). Guidelines revealing an appropriate amount of quantitative outputs would greatly facilitate simple cause–effect evaluation for different design scenarios. In this sense, efforts are still in great need to bridge the gap between science and practice at the relevant scales.

References

Ahmed KS (2003) Comfort in urban spaces: defining the boundaries of outdoor thermal comfort for the tropical urban environments. Energy Build 35:103–110

Ali-Toudert F, Mayer FH (2007) Effects of asymmetry, galleries, overhanging facades and vegetation on thermal comfort in urban street canyons. Sol Energy 81:742–754

Ali-Toudert F, Mayer H (2006) Numerical study on the effects of aspect ratio and orientation of an urban street canyon on outdoor thermal comfort in hot and dry climate. Build Environ 41:94–108

Allen L, Lindberg F, Grimmond CSB (2011) Global to city scale urban anthropogenic heat flux: model and variability. Int J Climatol 31(13):1990–2005

Arnfield AJ (2003) Review: Two decades of urban climate research: a review of turbulence, exchanges of energy and water, and the urban heat island. Int J Climatol 23:1–26

ASHRAE (2001) ASHRAE Fundamentals Handbook 2001 (SI Edition). Atlanta, USA, GA

Blocken B (2015) Computational fluid dynamics for urban physics: importance, scales, possibilities, limitations and ten tips and tricks towards accurate and reliable simulations. Build Environ 91:219–245

Bourbia F, Boucheriba F (2010) Impact of street design on urban microclimate for semi arid climate (Constantine). Renew Energy 35:343–347

Brown GZ, DeKay M (2001) Sun, wind & light: architectural design strategies, 2nd edn. Wiley, New York

Chatzidimitriou A, Yannas S (2015) Microclimate development in open urban spaces: the influence of form and materials. Energy Build 108:156–174

Chen Y, Wong NH (2006) Thermal benefits of city parks. Energy Build 38:105–120

Cheng V, Ng E (2006) Thermal comfort in urban open spaces for Hong Kong. Archit Sci Rev 49(3):236–242

Cheng V, Ng E, Chan C, Givoni B (2012) Outdoor thermal comfort study in sub-tropical climate: a longitudinal study based in Hong Kong. Int J Biometeorol 56(1):43–56. https://doi.org/10.1007/s00484-010-0396-z

Counihan J (1975) Adiabatic atmospheric boundary layers: a review and analysis of data from the period. Atmos Environ 9:871–905

Davenport AG, Grimmond CSB, Wieringa TROJ (2000) Estimating the roughness of cities and sheltered country. Paper presented at the 12th conference on applied climatology, Boston

Eliasson I (2000) The use of climate knowledge in urban planning. Landscape Urban Plan 48:31–44

Emmanuel R, Rosenlund H, Johansson E (2007) Urban shading—a design option for the tropics? A study in Colombo, Sri Lanka. Int J Climatol 27:1995–2004

Erell E, Pearlmutter D, Boneh D, Kutiel P (2014) Effect of high-albedo materials on pedestrian heat stress in urban street canyons. Urban Clim 10:367–386

Fanger PO (1972) Thermal comfort. Analysis and application in environment engineering. McGraw Hill Book Company, New York

Fukuoka Y (1997) Biometeorological studies on urban climate. Int J Biometeorol 40:54–57

Gagge AP, Fobelets AP, Berglund LG (1986) A standard predictive index of human response to the thermal environment. ASHRAE Trans 92(2B):709–731

Gaitani N, Mihalakakou G, Santamouris M (2007) On the use of bioclimatic architecture principles in order to improve thermal comfort conditions in outdoor spaces. Build Environ 42(1):317–324. https://doi.org/10.1016/j.buildenv.2005.08.018

Givoni B (1976) Man, climate and architecture. Applied Science Publishers, London

Givoni B (1994) Passive and low energy cooling of buildings. Van Nostrand Reinhold, New York

Givoni B, Noguchi M, Saaroni H, Pochter O, Yaacov Y, Feller N, Becker S (2003) Outdoor comfort research issues. Energy Build 35(1):77–86

Grimmond CSB (2006) Progress in measuring and observing the urban atmosphere. Theor Appl Climatol 84(1–3):3–22

Grimmond CSB, Oke T (1999) Aerodynamic properties of urban areas derived from analysis of surface form. J Appl Meteorol 38:1261–1292

Grimmond CSB, Roth M, Oke TR, Au YC, Best M, Betts R, Voogt J (2010) Climate and more sustainable cities: climate information for improved planning and management of cities (producers/capabilities perspective). Procedia Environ Sci 1:247–274

Höppe P (1984) Die Energiebilanz des Menschen (English translation: The energy balance in human), vol 49. Wiss Mitt Meteorol Inst Univ Mtinchen

Höppe P (1992) A new procedure to determine the mean radiant temperature outdoors (in German). Wetter und Leben, 44:147–151

Höppe P (1997) Aspects of human biometeorology in past, present and future. Int J Biometeorol 40:19–23

Höppe P (1999) The physiological equivalent temperature—a universal index for the biometeorological assessment of the thermal environment. Int J Biometeorol 43:71–75

Höppe P (2002) Different aspects of assessing indoor and outdoor thermal comfort. Energy Build 34:661–665

Hawkes D, McDonald J, Steemers K (2002) The selective environment. Spon Press, London; New York
Holst J, Mayer H (2011) Impacts of street design parameters on human-biometeorological variables. Meteorol Z 20(5):541–552
ISO (1994) Moderate thermal environments- determination of the PMV and PPD indices and the specification of conditions for thermal comfort. International Standard 7730, Geneva
Jauregui E (1997) The last Ms for 40th anniversary issue. Aspects of urban human biometeorology. Int J Biometeorol 40:58–61
Jones DL (1998) Architecture and the environment: bioclimatic building design. Overlook Press, Woodstock, NY
Kántor N, Unger J (2011) The most problematic variable in the course of human-biometeorological comfort assessment—the mean radiant temperature. Cent Eur J Geosci 3(1):90–110
Kanda M (2006) Progress in the scale modeling of urban climate: review. Theor Appl Climatol 84(1–3):23–33
Ketterer C, Matzarakis A (2014) Human-biometeorological assessment of heat stress reduction by replanning measures in Stuttgart, Germany. Landscape Urban Plann 122:78–88
Koch W (1962) Relationship between air temperature and mean radiant temperature in thermal comfort. Nature 196:587
Kwok AG, Grondzik WT (2007) The green studio handbook: environmental strategies for schematic design. Architectural Press, Oxford
Lai D, Zhou C, Huang J, Jiang Y, Long Z, Chen Q (2014) Outdoor space quality: a field study in an urban residential community in central China. Energy Build 68(B):713–720
Landsberg HE (1972) The assessment of human bioclimate, a limited review of physical parameters. Technical Note No. 123, WMO-No. 331. World Meteorological Organization, Geneva
Landsberg HE (1981) The urban climate. New York Academic Press
Lau KK-L, Lindberg F, Rayner D, Thorsson S (2015) The effect of urban geometry on mean radiant temperature under future climate change: a study of three European cities. Int J Biometeorol 59(7):799–814
Lau KK-L, Ren C, Ho J, Ng E (2016) Numerical modelling of mean radiant temperature in high-density sub-tropical urban environment. Energy Build 114:80–86
Lee H, Holst J, Mayer H (2013) Modification of human-biometeorologically significant radiant flux densities by shading as local method to mitigate heat stress in summer within urban street canyons. Adv Meteorol, Article ID 312572. https://doi.org/10.1155/2013/312572
Lee H, Mayer H, Chen L (2016) Contribution of trees and grasslands to the mitigation of human heat stress in a residential district of Freiburg, Southwest Germany. Landscape Urban Plann 148:37–50
Lin T-P (2009) Thermal perception, adaptation and attendance in a public square in hot and humid regions. Build Environ 44:2017–2026
Lindberg F, Grimmond CSB (2011) The influence of vegetation and building morphology on shadow patterns and mean radiant temperatures in urban areas: model development and evaluation. Theor Appl Climatol 105:311–323. https://doi.org/10.1007/s00704-010-0382-8
Lindberg F, Holmer B, Thorsson S (2008) SOLWEIG 1.0—Modelling spatial variations of 3D radiant fluxes and mean radiant temperature in complex urban settings. Int J Biometeorol 52(7):697–713
Lindberg F, Onomura S, Grimmond CSB (2016a) Influence of ground surface characteristics on the mean radiant temperature in urban areas. Int J Biometeorol. https://doi.org/10.1007/s00484-016-1135-x
Lindberg F, Thorsson S, Rayner D, Lau K (2016b) The impact of urban planning strategies on heat stress in a climate-change perspective. Sustain Cities Soc 25:1–12
Masmoudi S, Mazouz S (2004) Relation of geometry, vegetation and thermal comfort around buildings in urban settings, the case of hot arid regions. Energy Build 36(7):710–719
Matzarakis A, Mayer H, Iziomon MG (1999) Applications of a universal thermal index: physiological equivalent temperature. Int J Biometeorol 43:76–84
Mayer H (1993) Urban bioclimatology. Experientia 49(11):957–963

Mayer H, Höppe P (1987) Thermal comfort of man in different urban environments. Theor Appl Climatol 38:43–49
McGregor GR (2011) Human biometeorology. Prog Phys Geogr 36(1):93–109
McHarg IL (1992) Design with nature. Wiley, New York
Ng E, Yuan C, Chen L, Ren C, Fung JCH (2011) Improving the wind environment in high-density cities by understanding urban morphology and surface roughness: a study in Hong Kong. Landscape Urban Plann 101(1):59–74. https://doi.org/10.1016/j.landurbplan.2011.01.004
Nikolopoulou M (ed) (2004) Designing open spaces in the urban environment: a bioclimatic approach. Centre for Renewable Energy Sources, EESD, FP5, Athens
Nikolopoulou M, Baker N, Steemers K (2001) Thermal comfort in outdoor urban spaces: understanding the human parameter. Sol Energy 70:227–235
Oke TR (1981) Canyon geometry and the nocturnal heat island: comparison of scale model and field observations. J Climatol 1:237–254
Oke TR (1987) Boundary layer climates, 2nd edn. Methuen, London
Oke TR (1988) Street design and urban canopy layer climate. Energy Build 11(1–3):103–113
Oke TR (2006a) Initial guidance to obtain representative meteorological observations at urban sites. World Meteorological Organization, Instruments and Observing Methods, IOM Report No. 81, WMO/TD-No. 1250
Oke TR (2006b) Towards better scientific communication in urban climate. Theor Appl Climatol 84:179–190
Olgyay V (1963) Design with climate: bioclimatic approach to architectural regionalism. Princeton University Press, Princeton, NJ
Pickup J, de Dear R (1999, 8–12, November) An outdoor thermal comfort index (OUT_SET*)—Part 1 The model and its assumptions. Paper presented at the 15th International Congress of Biometeorology & International Conference of Urban Climatology, Sydney, Australia
Radhi H, Fikry F, Sharples S (2013) Impacts of urbanisation on the thermal behaviour of new built up environments: a scoping study of the urban heat island in Bahrain. Landscape Urban Plann 113:47–61
Rotach MW, Vogt R, Bernhofer C, Batchvarova E, Christen A, Clappier A, Voogt JA (2005) BUBBLE—an urban boundary layer meteorology project. Theor Appl Climatol 81:231–261
Sargent FI (1965) A survey of human biometeorology. Int J Biometeorol 9(1):1–3
Sargent FI, Tromp S (1966) The first decade of the International Society of Biometeorology (1956–66). Int J Biometeorol 10:207–214
Souch C, Grimmond S (2006) Applied climatology: urban climate. Prog Phys Geogr 30(2):270–279
Spagnolo J, De Dear R (2003) A field study of thermal comfort in outdoor and semi-outdoor environments in subtropical Sydney Australia. Build Environ 38(5):721–738
Stathopoulos T (2006) Pedestrian level winds and outdoor human comfort. J Wind Eng Ind Aerodyn 94(11):769–780
Swaid H, Bar-El M, Hoffman ME (1993) A bioclimatic design methodology for urban outdoor spaces. Theor Appl Climatol 48(1):49–61
Szokolay SV (2004) Introduction to architectural science: the basis of sustainable design. Architectural Press, Oxford
Takahashi K, Yoshida H, Tanaka Y, Aotake N, Wang F (2004) Measurement of thermal environment in Kyoto city and its prediction by CFD simulation. Energy Build 36:771–779
Task Committee on Outdoor Human Comfort of the Aerodynamics, Committee of the American Society of Civil Engineers (2004) Outdoor human comfort and its assessment: state of the art. American Society of Civil Engineers, Reston, VA
Thorsson S, Lindqvist M, Lindqvist S (2004) Thermal bioclimatic conditions and patterns of behaviour in an urban park in Goteborg. Int J Biometeorol 48:149–156
Thorsson S, Rocklöv J, Konarska J, Lindberg F, Holmer B, Dousset B, Rayne D (2014) Mean radiant temperature—a predictor of heat related mortality. Urban Clim 10(2):332–345

Tseliou A, Tsiros IX, Lykoudis S, Nikolopoulou M (2009) An evaluation of three biometeorological indices for human thermal comfort in urban outdoor areas under real climatic conditions. Build Environ 45(5):1346–1352

Vanos JK, Warland JS, Gillespie TJ, Kenny NA (2010) Review of the physiology of human thermal comfort while exercising in urban landscapes and implications for bioclimatic design. Int J Biometeorol 54(4):319–334. https://doi.org/10.1007/s00484-010-0301-9

VDI 3787 (1996) Richtlinie VDI 3787: Methoden zur human-biometeorologischen Bewertung von Klima und Lufthygiene für die Stadt- und Regionalplanung. Teil 1: Klima. (English translation: VDI 3787: Methods for the human biometeorological assessment of climate and air hygiene for urban and regional planning Part 1: Climate). Technical report: Association of German Engineers

Yang F, Chen L (2016) Developing a thermal atlas for climate-responsive urban design based on empirical modeling and urban morphological analysis. Energy Build 111:120–130

Yang S-R, Lin T-P (2016) An integrated outdoor spaces design procedure to relieve heat stress in hot and humid regions. Build Environ 99:149–160

Chapter 3
Asian High-Rise Urbanism

Abstract This chapter firstly looks back in retrospect at the brief history and precedents of high-rise urbanism and then reviews the contemporary development of high-rise urbanism in Asian megacities by a comparison in economic, social, and environmental aspects of three Asian megacities, Singapore, Hong Kong, and Shanghai. The comparative study, with respect to climate, geography/topography, demography, urban governance and policy, etc., reveals that, while the new town system featuring high-rise high-density neighborhood and public transit-oriented development, as adopted by Singapore and Hong Kong, has effectively responded to the critical challenges such as land restriction in the progression of urbanization and evolved toward sustainable compact urban forms, the city of Shanghai seems to be influenced by the new urbanism movement mostly found in North American cities. With largely unsatisfactory government support in urban public service infrastructure and land-use planning for compact development, the intended city center redevelopment and suburbanization (the One City and Nine Towns scheme) seem to have led toward urban sprawling and deteriorating urban environment. The comparison indicates a more compact high-rise urban form may better support environmental sustainable development of Shanghai, and an integrated climate-responsive planning and design workflow are to be developed, and better inter- and cross-disciplinary communication is needed between urban climatology and urban design. For that, a framework is proposed for microclimate research and design integration for high-rise urbanism, and the parametrization of urban form on microclimate at three scales is listed for the followed case studies in Part Two.

3.1 High-Rise Urbanism in Asian Megacities

3.1.1 Precedents of High-Rise Urbanism

High-rise was emerged as a building typology for office and commercial space in North American cities around the late nineteenth century. Driven by soaring land price in the city center and supported by technological advancement in steel frame construction and modern elevator systems, tall buildings with tens of floors high

F. Yang and L. Chen, *High-Rise Urban Form and Microclimate*,
The Urban Book Series, https://doi.org/10.1007/978-981-15-1714-3_3

appeared and blossomed in center area of cities such as Chicago and NYC. High-rise building as a key instrument for a new urbanism, however, was probably firstly advocated by the Swiss architect Le Corbusier (Charles-Edouard Jeanneret), in his (then) bold proposal of the *Contemporary City* for 3 million inhabitants (Corbusier 1987).

The proposal has been criticized for its ignorance of traditional urban fabric and separation of residence, work, and entertainment functions in urban zoning. Le Corbusier himself admits that "*these plans are tentative, and they do not take into account the compromises which would be necessary.*" Nevertheless, what the plans strive for is very forward-looking at the time, i.e., every citizen's health, comfort, and well-being in working, dwelling, and moving around in their city. To achieve these, the architect advocated that the new urbanism should embrace the Machine Age and leave all the nostalgia for historic towns behind. He claimed that a city of machine-like is needed: A city center comprised of 60-story high skyscrapers and surrounded by a number of *garden cities*, connected with a vertically layered transportation system (i.e., in the order from bottom-up: underground tubes and heavy cargo traffic ordinary street-elevated fast arterial road). Green spaces are considered indispensable by Le Corbusier as he calls them *the lungs*. Skyscrapers are therefore arrayed in large parks and green open spaces, with sufficient distances from each other so as to ensure required standards of daylighting and irradiation, herein lies the idea of *towers in the park*. In the center of the skyscraper cluster located the main transportation deck. The major component of urban housing is 12–15-story apartment buildings known as *Unités*. The amenities and services are inserted into the vertical stack of cellular flats: catering and laundry at the bottom and a kindergarten and a pool on the roof. This idea of so-called *vertical village* is to reduce daily commuting and encourage social activities within the building footprint. The Unité d'Habitation Marseille built in 1947 and designed by Le Corbusier is a vivid interpretation of the idea.

Building tall, as can be seen in Le Corbusier's grandiose plan, is a key to healthy and comfortable modern urban living. Building tall and living high is not necessarily a problem. Le Corbusier carefully considered and specifically regulated density (floor area ratio and population per acre) and open space ratios for working and residential areas in city center of his Radiant City Scheme. Only by building tall can he acquire the desired density, meet the daylight and irradiation control targets as well as release large portions of ground area for green open space. The spirit in the Radiant City Scheme has deep influences on planning and housing design practice ever since. For instance, the concept of vertical village has inspired the idea of Ecological Skyscraper advocated by Kenneth Yeang and works of renowned architects, e.g., the Linked Hybrid in Beijing (by Steven Holl Architects). It should be noted that urban housing blocks have to be considered a component connected with the complex urban system. Many factors contribute to the success or failure of high-rise apartment blocks, for instance, location, estate management, and social/economic status of occupants. While some suburban blocks made of monolithic towers were obsoleted due to poverty and lack of sense of community, there are many cases of successful *towers in the park* scheme in urban housing projects in North American cities, for instance, the Stuyvesant Town and Peter Cooper Village ("Study Town") and Penn South in Manhattan, and Co-op City in Bronx NYC, among others. Not to mention that, the high-rise apartment blocks (quarters) development is becoming the dominant housing typology in Asian megacities such as Singapore, Hong Kong, and Shanghai.

3.1.2 *High-Rise Urbanism in Singapore, Hong Kong, and Shanghai*

The three cities (or city-state) are chosen for comparison because that, they are among the most developed in the Asian megacities. Shanghai is the largest and most populous city in China. Although the GDP per capita of Shanghai is not the highest (as of in 2018) in China, the overall GDP totaled 3.27 trillion yuan (480 billion dollars) in 2018, on top of all Chinese cities. Hong Kong is a special administrative region of China. Chinese is the largest ethnic group in Singapore population. Therefore, the three cities share some cultural roots. However, there are still clear differences among the three cities in terms of climate, geography/topography, and urban development (Table 3.1).

- Singapore

Singapore and Hong Kong are both situated on island/peninsula and thus limited in developable land area. Singapore, as a city-state, has chosen high-rise and high-density pattern for urban development ever since the beginning of its independence,

Table 3.1 A comparison of three Asian megacities: Singapore, Hong Kong, and Shanghai

	Singapore	Hong Kong	Shanghai
Latitude	1° N	22.3° N	31.2° N
Climate (Koppen)	Af-tropical, hot, and humid	Cmb-warm and humid, monsoonal	Cfa-hot summer humid all year; monsoonal
Land area (km^2)	722.5	1108	663.7 (city area)
GDP per capita (USD)	55,000	38,000	20,000
Population (million)	561	735	1116 (city area)
Housing scheme	80% residents in HDB public housing, ~90% with homeownership	50% residents in subsidized public housing rental + owner	Private housing, with negligible portion of public housing
Urban development	New town system	New town system	New urbanism
Housing FAR range	4–9	5–7.5	2–2.5
Housing floor no. (mean)	40–50F	30–50F	>9F (27 m)

to accommodate the foreseeable long-term increase in population. The high-rise pattern makes it possible to preserve valuable land for future development and to preserve natural environment: Currently, more than one-third of the Singapore Island is reserved for natural green space and covered with vivid vegetation. The government's slogan is to build Singapore into a *Garden City*.

More than 80% of Singaporeans live in public housing developed by the Housing and Development Board (HDB), the public housing authority of Singapore. The development follows the new town urbanism. There are 23 towns and 3 estates across the island. Each town has a clear hierarchical structure: the Town (150,000–250,000 people)–Neighborhood (20,000–30,000 people, in 4000–6000 households)–Precinct (400–800 households). Public services and commercial/amenity spaces of various scales are allocated at each level of the structure accordingly. The first generation of public housing blocks are 20–30 stories and are being replaced by HDB with significantly higher towers of around 50 stories. And modular and standardized housing designs are being shifted toward customized and non-standardized architectural design. An example is The *Pinnacle@Duxton*. The project is recognized for its innovative design in providing vivid urban life in an ultra-high density (Fig. 3.1).

The Pinnacle@Duxton was completed and put into use in 2009. It comprises seven apartment towers of 50 stories high, arranged in a curvilinear layout, and connected with each other at the 3rd, 26th, and 50th floors. These connected floor areas are used for communal amenity facilities and green spaces, contributing to the sense of community, and resonating in a way with the idea of "vertical village" coined by Le Corbusier a hundred years ago. With a FAR of 9.3, the project provides 1848 flats and accommodates nearly 7000 residents with a site of 2.5 ha. This means a density of nearly 3000 people per hectare, a density equal to the city center area of the Radiant City Scheme by Le Corbusier, and ten times higher than that of the residential area of the Scheme. Despite its modular construction using replicated prefabricated elements, the building façade has somehow obtained an alternating pattern with limited types of prefab-façade panels, therefore, reduced the perceived building mass and improved the, otherwise, monolithic appearance.

Under a tropical climate and built in such a high density, urban climate, and human thermal comfort has been an important research topic. The major concern is to alleviate heat stress by reducing solar gain, enhancing airflow, and mitigate UHI effect. Studies have been carried out on urban heat islands, urban greenery, building morphology, and energy (Wong and Chen 2009). For instance, a microclimate and UHI evaluation tool are developed to estimate pedestrian-level comfort and air temperature differential based on urban morphology parameters associated with building, pavement, and greenery (Jusuf and Wong 2009).

- Hong Kong

Hong Kong has a hilly terrain, and over three quarters of the total 1100 km^2 land area is not suitable for development. It seems that Hong Kong has no other options but to go high-rise and high density. Tall buildings packed in confined urban areas have been spectacular urban scenes for visitors and tourists to the City of Oriental Pearl.

(a)

(b)

Fig. 3.1 Public housing project The Pinnacle@Duxton, Singapore (**a**). Communal amenity spaces are located on the raised new ground, 26th and 50th stories, respectively (**b**). *Photograph credit* **a** Liu Te, **b** Chen Qian. Used with permission

Like Singapore, the new town development had been an essential way of Hong Kong in its progression of urbanization, to cope with the increase in population and to improve the living environment by decentralizing the population from the overcrowded city center. The basic concept for developing a new town is to provide a balanced and self-contained community as far as possible in terms of the provision (or upgrade) of infrastructure and community facilities. Hong Kong has developed nine new towns since the initiation of its New Town Program in 1973. The nine new towns could be divided into three generations. The first three are Tsuen Wan, Sha Tin, and Tuen Mun, which started works in the early 1970s, when the urban area in Hong Kong was still relatively small, restricted to the central and western part of Kowloon Peninsula and the northern side of Hong Kong Island; then the second three, Tai Po, Fanling/Sheung Shui, and Yuen Long in the late 1970s; and the third three, Tseung Kwan O, Tin Shui Wai, and Tung Chung in the 1980s and 1990s. The population of these nine new towns is about 3.47 million (data of May 2016) and is expected to rise to 3.63 million in 2021.

Public housing comprises a significant part of the residential building sector in Hong Kong. Nearly a half of Hong Kong people live in it (www.censtatd.gov.hk). Public housing is developed by Hong Kong Housing Authority (HKHA) and designed and built by the Housing Department. Through tens of years of development, the public housing of Hong Kong has evolved into a system of standardized design with respect to housing unit layout and its combination (e.g., the Harmony Model 和谐式 in current use). The standardization facilitates prefabrication and cost control in the construction process. But it can be less flexible in responding to local terrain/topography and microclimate.

The high-density high-rise urban form of Hong Kong has given rise to a series of environmental problems such as stagnant urban ventilation, deteriorated street-level air quality, and elevated UHI effects. After the SARS epidemic break out in 2003, the Hong Kong government finally realized the seriousness of the problems and initiated comprehensive assessment studies on urban air ventilation and urban climate mapping (Ng 2008). Thermal loads and dynamic potential are quantified using urban morphological parameters, e.g., sky view factor, frontal area index. (Planning Department Hong Kong SAR 2009). These studies provide valuable information, knowledge, and guidance to authorities and industries. Following these, the *Air Ventilation Assessment (AVA) in Chapter 11, Urban Design Guidelines,* of the Hong Kong Planning and Design Guidelines was issued by Planning Department (www.pland.gov.hk), and *Sustainable Building Design Guidelines* was issued by Building Department of Hong Kong (APP 151, 152) (www.bd.gov.hk). These guidelines stipulate key aspects in building separation, set back and green space coverage so as to tackle with the urban environmental problems in air ventilation, thermal comfort, and pollutant dispersion (Yuan and Ng 2012).

For Hong Kong, a question to be answered would be, how dense and how high are the limits? It seems that Hong Kong is continuing challenging the extreme in the urbanization progress. The current residential development is in average 60 stories, and even higher standards are planned for the future. To make the high-density living more sustainable, it'd be vital to develop an integrated environmental-responsive planning and design workflow. An example is the redevelopment of *Upper Ngau Tau Kok Estate (UNTK) Phase 2 and Phase 3* (Fig. 3.2a). Completed in 2009, the project comprises six 40-story high apartment buildings in a 3.2-ha site, providing 4584 units with a total building floor area of 220,000 m^2. The FAR is nearly 7. Modular flat design is used to facilitate prefabrication and expedite construction, just as the past generations of public housing in HK. But unlike its precedents, which standardize site planning by arrayed identical residential towers, UNTK carefully organizes the buildings to improve the quality and usability of the ground-level public space, indoor/outdoor ventilation, thermal comfort, daylighting and acoustic environment (Fig. 3.2b). In particular, to quantitatively assess the site ventilation, numerical simulation based on computational fluid dynamics (CFD) was employed to assist in the planning and building design (Ng 2010). The project is accredited and awarded the 3-star (the highest) certificate of the China Green Building Design Label.

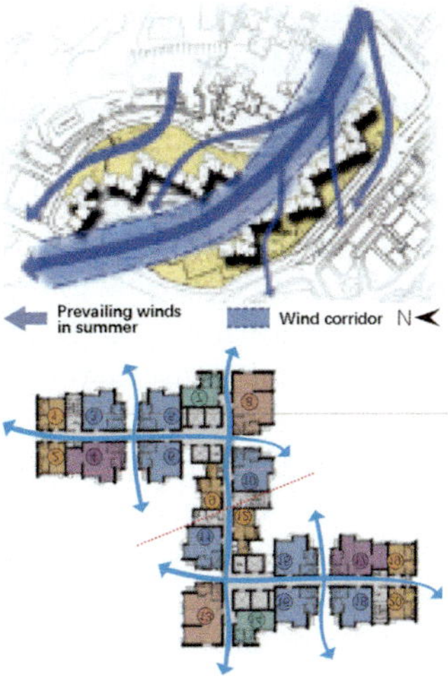

Fig. 3.2 Upper Ngau Tau Kok Estate Phase 2 and Phase 3. *Source* Fung et al. 2011 used with permission

- Shanghai

The metropolitan Shanghai, including city center, urban, and suburban areas, is 6340.5 km^2. But the central city area, bounded with the outer ring highway, is 664 km^2 and accommodates more than 11 million people, nearly a half of overall population of Shanghai. Compared to Singapore, central city of Shanghai holds about one time more people with a smaller land area (Table 3.1). On the other hand, Singapore has an average FAR of public housing development around 7, whereas Shanghai strictly regulates density of residential development in urban areas: FAR and density in the inner-city area (bounded by the inner-ring highway) are no higher than 2.5 and 25%, respectively; and in-between the inner and outer rings no higher than 2 and 25%, respectively (Shanghai Government 2003). Accessibility to mass rail transit system has an effect on FAR control and allocation: The site in close proximity to MRT stations may be allowed some extra volume of salable floor area but the threshold is quite high and the reward is not generous (Shanghai Bureau of Planning and Land Resources 2016). The intent may be to lower the housing density in city center so as to improve the living condition, but the similar development intensity in city center and its surrounding area may not be the most optimized use of urban land.

Shanghai has been actively urbanizing its vast suburban area, transmitting urban dwellers from dense city center to suburban new towns, and redevelop the obsoleted traditional city areas. Started in 2001, Shanghai government initiated the so-called One City and Nine Towns pilot scheme of a New Town-making movement. These are in fact ten new towns surrounding the central city area: a large and relatively developed one, Songjiang New Town (Songjiang was the ancient town center of Shanghai) and nine new towns which are the central municipal towns of nine suburban districts (Anting of Jiading District, Haigang of Nanhui District, Gaoqiao of Pudong

New District, Fengjing of Jinshan District, Chenjianzhen of Chongming County, Pujiang of Minhang District, Fengcheng of Fengxian District, Luodian of Baoshan District, and Zhujiajiao of Qingpu District. Nanhui was merged into Pudong New District in 2009). The pilot scheme is criticized for its theme-park style development: Each new town was assigned a Western urban and architecture style, and Western architects were invited to bid for planning and building design. For instance, Anting is the German-style new town. On the other hand, the key principles of new town urbanization that has been successfully applied in Western countries as well as in Singapore and Hong Kong were not carefully studied and implemented. Planned development density is low (FAR < 1.8 for new town and <1 for other rural district areas). Job and employment often cannot be contained locally, and a large amount of new town residents still have to work in central city area. For commuting, mass rail transit system largely has insufficient capacity, and new town people have to rely heavily on individual vehicle transportation. The result is the new towns as "sleeping towns," with inconvenient connections to central city area (顾竹屹 et al. 2014).

The new town system of Singapore contributes to a compact urban form with mixed land uses and building functions. Despite the limited land area, about one-third of Singapore is reserved as green space which contributes to the State Image of Garden City. As for Shanghai, the green space ratio is 21%. But this ratio is based on the metropolitan area and includes the vast suburban area, whereas the green space ratio of the central city area can be significantly lower. With less compact urban form, low development intensity and insufficient green space, the urbanization of Shanghai is more like the New Urbanism of North American cities instead of the new town system in Hong Kong and Singapore. The results are observed as urban sprawling with generic buildings and featureless urban spaces. This does not meet the compact urban form for sustainable development.

With aspect to geography, Shanghai has its vast hinterland, thus less restrictions in land reserve. With respect to climate, the colder winter of Shanghai compared to Hong Kong and Singapore leads to wider spacing between residential buildings to ensure certain solar irradiation in heating season, and this consequently lowers housing density. An investigation on FAR potential of Shanghai new towns took into account building spacing for irradiation, service radius of neighborhood amenity center and transportation capacity and concludes that depending on residential building type and height limit, the FAR of new town should not be higher than 2.2 for the 60-m-height limit (Xu 2016). The premises in winter solar irradiation calculation is very generous, i.e., 4 h' continuous sunshine in winter solstice day compared to 1-h irradiation regulated by the national standard on urban residential development. Also, the transportation capacity calculation assumes a quite high percentage of private vehicle traffic, and therefore, more land used for driveway. It is possible to go for higher density than this, provided that urban infrastructure planning can support better public transportation and service center capacity, while reducing reasonably over high requirement on winter solar radiation.

Optimization in density has to be made by trade-off and compromises between economic (e.g., land price, construction cost), technical (e.g., construction, fire regulation, and building law: building higher than 100 m is classified super-tall and cost increased significantly due to higher standard on construction and more strict fire protection, and egress regulation), social (e.g., residents' perception and acceptance of density) and environmental (e.g., daylight, sun, wind, comfort) factors. For instance, housing price can be negatively correlated with housing density (Fesselmeyer and Seah 2018). In Hong Kong, the public housing sector accommodates nearly half of urban population while in Singapore more than 80%. The cost of public housing (either rental or ownership) is much lower than the private housing sector. Therefore, the elevated density that comes with the cost benefit seems to be more acceptable (or tolerable) by public housing residents. Whereas in Shanghai, as well as other Chinese cities, urban housing has been transformed into market-driven sector ever since the beginning of the century, and private housing comprises the absolute majority of the residential sector. It may be well understood by the government as well as developers that to challenge the boundary of compact development by testing higher housing density will put the commercial interests and land/tax revenue at risk. So the invisible hand of the housing market economy has indirectly contributed to a lower housing density and the relatively less challenging physical environment.

According to the strategic planning of the government, subsidized rental housing is to be increasingly supplied in near future in Shanghai. With economic and social factors considered, density of this housing type should be elevated compared to commercial housing. Consequently, the density will pose challenges to the indoor and outdoor environment and microclimate, and it is important to develop integrated climate-responsive design strategies.

3.2 Microclimate Research for High-Rise: A Framework

3.2.1 Communicating Urban Climatology with Urban Design

Microclimate research in high-rise high-density Asian cities such as Hong Kong and Singapore invariably take on an integrated approach to design evaluation for decision-support purposes, taking into account design parameters in building, land use, and greenery. Winter wind and comfort may be an issue for Shanghai but other than that, experiences gained and lessons learned from Hong Kong and Singapore are very valuable. In addition, mutual understanding between the "demand" and "supply" side in climate-responsive urban design as an interdisciplinary practice will help both sides in collaboration with better results.

To effectively inform urban design, urban climate research needs to overcome many barriers. First of all, the climate is one of the many issues to be considered in planning and design process, such as economics, culture, politics, technology, aesthetics, and very often the microclimate issue is not prioritized, but to be handled with or repaired later, by applying "supplementary" measures. For instance, with respect to the street aspect ratio (H/W) of a specific climate type, research took into account solar irradiation, shelter, heat island mitigation, etc., and proposes an ideal H/W range (Oke 1988). However, in real city center and particularly in megacities, development density and building height are determined by the economic factor (cost of land and return on investment), whereas road width is determined by vehicle traffic density planning and calculation. In addition, planners and architects may be reluctant to actively use climate information because of insufficient knowledge, unfamiliarity with the tools, worry about the added cost, and uncertainty about market and policy. (Eliasson 2000). Therefore, climate analysis and information cannot predetermine the modern urban form, as it may have done so in traditional vernacular towns, despite it is being widely acknowledged that the climate information should be integrated from the beginning of contemporary urban design process.

On the other hand, urban planning and design have long included urban climate as a part of the integrated framework, and urban design research also covers urban climate issues, but in ways quite differently with urban climatology (Table 3.2). Considering these, communication between the two can be vital in better integrating climate knowledge in urban design. Based on his experience and expertise, Edward Ng, Professor of CUHK and team leader of the AVA and urban climate study of Hong Kong, summarized that the key for urban climatologists is to keep report *simple and visual* and highlight what is "*prevailing*" and "*critical,*" i.e., how often the situation occurs and how important the problem is (to comfort, energy, and well-being) (Ng 2012). It is also important to use parameters that are related to urban form and land use, preferably directly applicable in planning and building design by planner/architect in their daily work, such as FAR, and avoid those professional parameters such as roughness length (z_0) and zero-plane displacement (Z_d) height. Based on these, data visualization using innovative digital technology should be of great help to communicate with and inform designers in understanding and assessing the design scenarios in terms of climate-responsive performance.

Table 3.2 A comparison of urban design and urban climatology (inspired by Erell (2008), Mills et al. (2010))

	Urban climatology	Urban design
Research focus	General phenomenon or climatic parameter, heat and mass transfer, UHI, pollutant concentrations, etc.	Specific case-based study, on the correlation of urban from parameters and microclimate, potential of design integration
Objective	Abstract, ideal urban geometry unit, e.g., street canyon, urban grid governed by geometric aspect ratio; or aerodynamic parameters, e.g., roughness length, etc.	Real-world cities, traditional or modern habitats and the people live in, e.g., vernacular settlements or contemporary urban design examples
Methodology	Quantitative: instrumented measurement or numerical simulation	Qualitative or quantitative: observation, investigation, survey, and interview
Presentation	Formulae and equations express the processes governed by thermos-dynamic principles, with assumptions and various degrees of simplification	Report by texts, photographs, and diagrams with support of survey data. Narrative describing the interaction of form and climate, and human's perception

3.2.2 *Parametrizing High-Rise Urban Form*

The urban geometry, fabric, and land cover have profound and synthetic effects on urban energy balance and microclimate (Fig. 3.3). A comprehensive and detailed review on urban microclimate is presented in Chap. 2. The review here focuses on high-rise and (sub) tropical urban and climate contexts of Asian megacities and solar control and ventilation at the pedestrian, block and neighborhood scale. The review is the basis of urban form parametrization that follows in this section.

In the four aspects of modification by urbanization, high-rise urban form strongly modifies the building volume/layout by providing a distinct third vertical dimension to the urban substrate surfaces in a magnitude that is not negligible. The high-rise urban configuration generates urban space with various degrees of geometrical and morphological complexity, depending on building materials and layout, spacing, height variation, and placement of void (porosity). The complex form has profound impact on the daily and seasonal energy balance and varying thermal comfort effects on urban spaces, with respect to solar heat gain and airflow. To assess the environmental performance of urban form for climate-responsive urban design, parameters at various scales need to be derived to quantify the effects of high-rise urban morphology and identify those that are sensitive to the interaction between form and the forces of sun and wind. In addition, parameters on urban cover (pavement and pervious surfaces) and urban metabolism (anthropogenic heat) are also to be included.

Radiant environment is associated with passive solar design of urban buildings and outdoor spaces. Thermal comfort effect of radiation can be measured by mean radiant temperature (T_{mrt}). In summer days under subtropical climates, T_{mrt} is very important and oftentimes the determining factor of outdoor comfort. To quantify the complex interplay of high-rise geometry and solar geometry, a parameter called **total site factor (TSF)** is introduced which, initially developed for forestry study, is

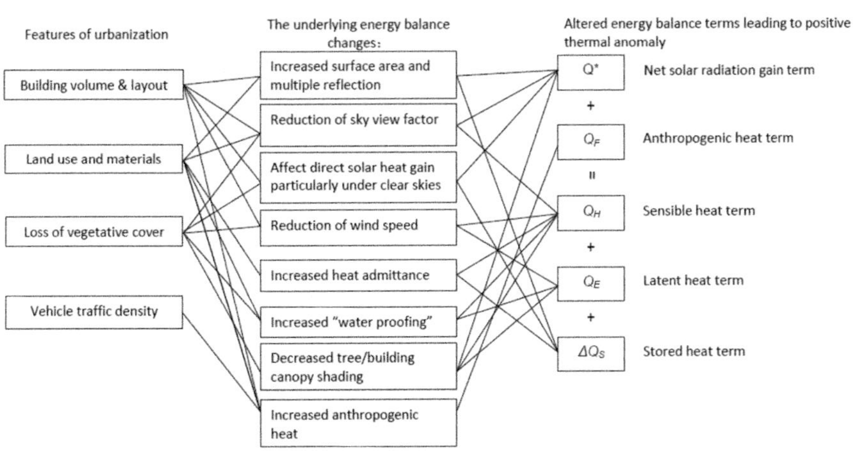

Fig. 3.3 Interconnection among urbanization and energy balance (after Oke (1987), pp. 275, 294)

basically the ratio of solar radiation under an urban canopy to that above the canopy (Table 3.3). **Sky view factor (SVF)** has been correlated negatively with the maximum urban heat island intensity and positively with the diurnal temperature fluctuations, as it indicates the capability of daytime shortwave heating and nighttime longwave cooling at ground horizontal surfaces (Oke 1981; Steemers and Steane 2004). It may also be related to daytime air movement caused by thermal buoyancy effect. The degree of openness (to the sky) can be an useful indicator of ventilation potential, in that the "degree of enclosure" created by building and/or vegetation canopy can be related to the probability of air stagnation.

For thermal comfort and pollutant dispersion, ventilation needs to be enhanced in high-rise urban environment. High-rise morphology in general produces stronger turbulence and provides better ventilation to ground area compared to low-rise morphology (Kubota et al. 2008). In high-rise high-density urban spaces, turbulence can be a major driver for air movement: overall wind speed is significantly lowered but depending upon the prevailing wind profile and surrounding urban geometry, the distribution of airflow is highly varied both spatially and temporally (Givoni 1998). To enhance air circulation, *horizontally*, density (building footprint) is considered a determining factor. For a given range of density, ground/figure configuration (refer to the famous Nori Map) should integrate *air corridor* with urban road and open space system; *vertically*, the degree of *porosity* (ratio of solid to void and continuity of void with respect to advection flow) is to be optimized to enhance cross-ventilation. A reasonable *variation in building height*, with respect to building spacing, is needed to increase surface roughness and turbulence production.

Based on the above understanding, to enhance the pedestrian-level airflow within the urban canopy layer (UCL), the production of turbulence should be enhanced above the roof level, i.e., in the urban boundary layer (UBL) (Oke 1988). **Building coverage ratio (BCR)** is an important indicator of surface roughness. Within the range of 0.13–0.32, increase in BCR firstly tends to increase the turbulence production, indicated by normalized z_0 and Z_d, but then decreases when it exceeds around 0.25 (Grimmond and Oke 1998). Surprisingly, the Shanghai urban planning regulation on residential development density perfectly meets the 25% threshold (Shanghai Government 2003) and thus encourages optimum turbulence production from the urban ventilation perspective. But the control on non-residential development density is less strict and thus may need case-based ventilation study. Tall and slim towers (dubbed "pencil buildings" in Hong Kong) are good for ventilation, as this form reduces the wall effect of aligned tall buildings on ventilation and pollutant dispersion (Yim et al. 2009). Features such as arcade (Qi Lou) also help improve pedestrian-level airflow and air change rate (Wen et al. 2017; Juan et al. 2017). For cities in monsoon climates, building clusters are organized by gradually lowering allowable building heights toward the summertime prevailing wind direction (Givoni 1998). Towers without protruding podiums can introduce more airflow from above down to the pedestrian area and recessed entrances, and portico façades can be ventilation-friendly features for pedestrians.

Table 3.3 List of urban morphology parameters

	Independent variables (abbreviations)	Schematic diagram	Measured environmental properties
Micro (pedestrian) scale	Sky view factor (SVF)	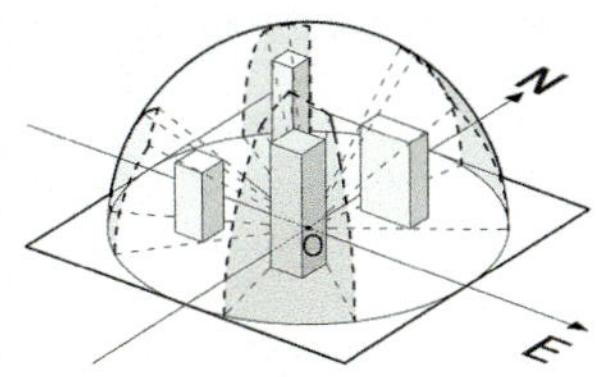	Defined as the ratio between the radiation received by a planar surface from the sky to the radiation emitted to the entire hemispheric radiating environment
	Total site factor (TSF)	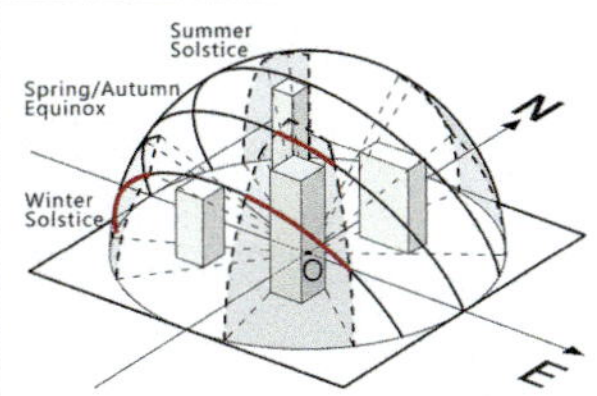	Solar admittance (integrated measured): the ratio of solar radiation (direct and diffuse) under the canopy to that above the canopy
	Tree view factor (TVF)	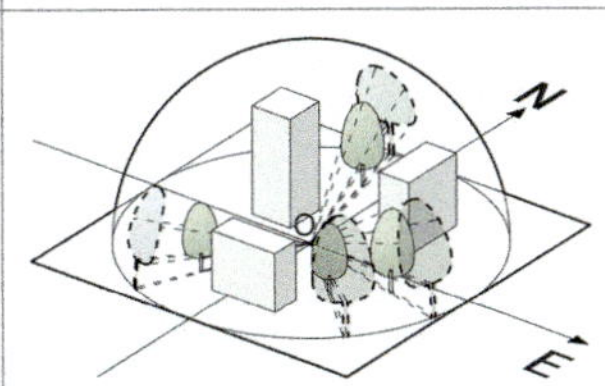	Shading factor by tree canopy: the fractions of the overlying hemisphere as shown in a sky view image that is occupied by vegetation canopy
	Green plot ratio (GPR)	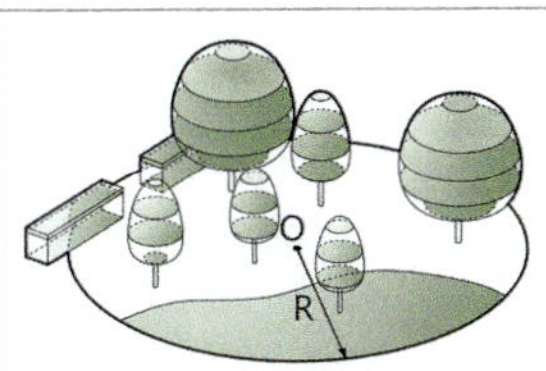	Vegetation leaf density (greenery mass): the ratio of total leaf area to a fixed-radius area centered by the point under measurement
	Green cover ratio (GCR)	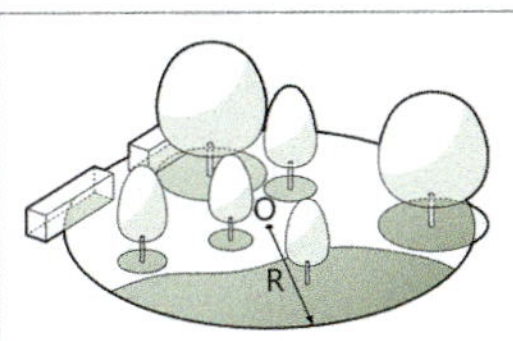	Ground greenery coverage at the micro scale: percentage of the projected area of all types of vegetation (grass, shrub, and tree) within a fixed-radius area of a measurement point
	Ground surface albedo (GSA)	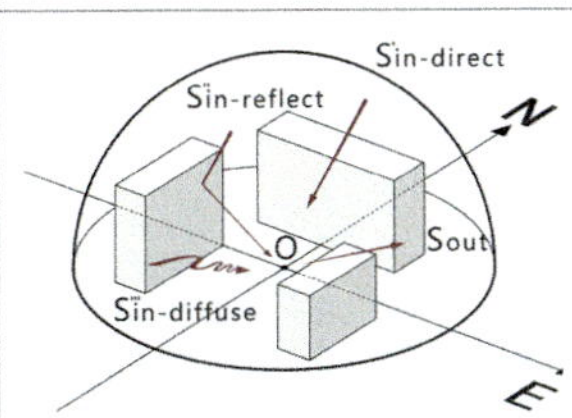	Defined as hemispherically and wavelength-integrated reflectivity of ground surfaces of measurement points

(continued)

Table 3.3 (continued)

	Independent variables (abbreviations)	Schematic diagram	Measured environmental properties
Block (site) scale	Building cover ratio (BCR)	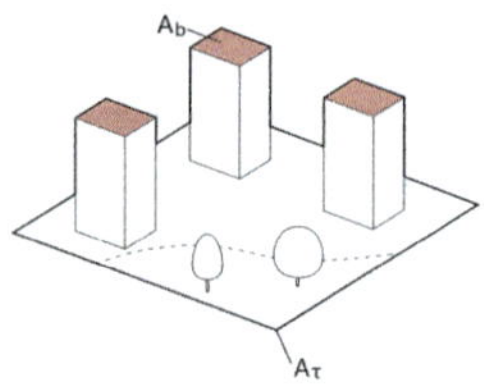	Ratio between building footprint and site area. Also known as building density in architecture and urban planning
	Hard pavement ratio (HPR)	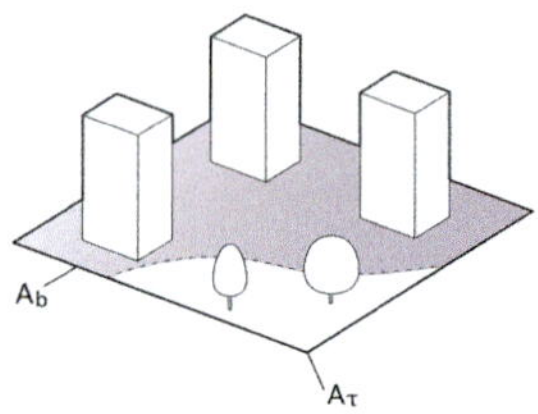	Ratio of total impervious pavement area to the site area
	Green ratio (GR)	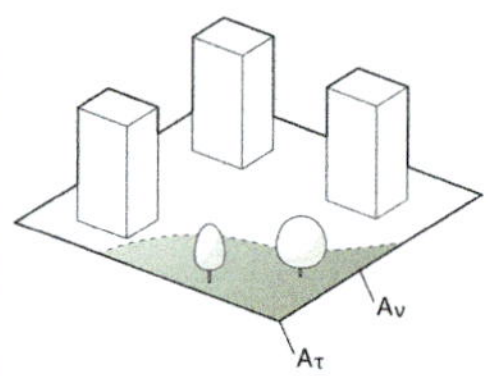	Ground greenery coverage on the site scale
	Frontal area density (FAD)	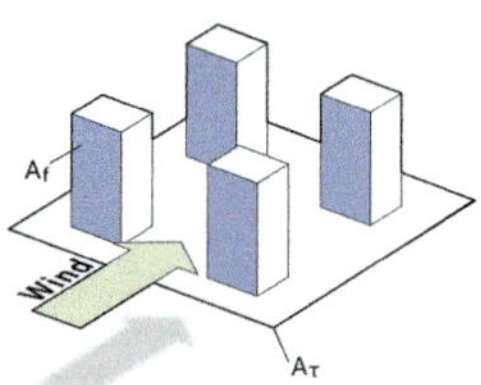	A measure of the frontal area per unit horizontal area per unit height increment, it quantifies the drag force as a function of building height: ventilation potential (Burian 2002)
	Floor area ratio (FAR)	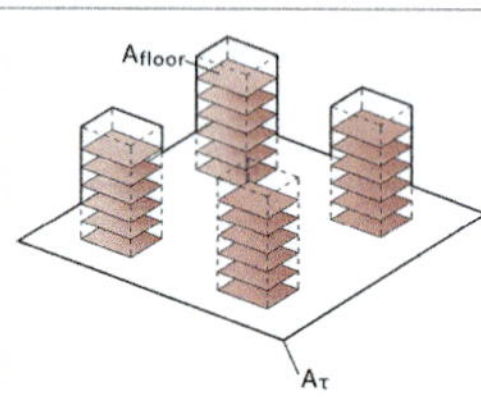	Plot building volumetric density: ratio of the total building floor area to the plot or site area
	Proximity to heat sink (PHS)	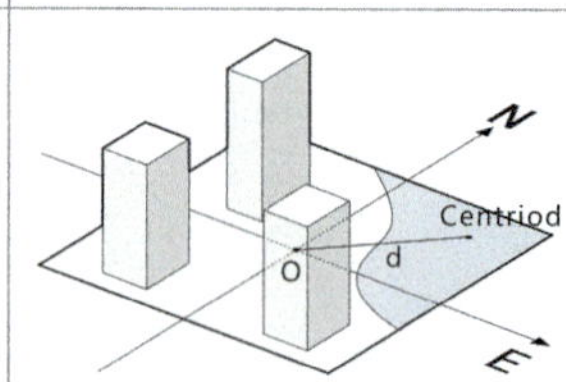	The distance to the centroid of any nearby sizable water body and green space

(continued)

Table 3.3 (continued)

	Independent variables (abbreviations)	Schematic diagram	Measured environmental properties
	Vehicle traffic density (VTD)	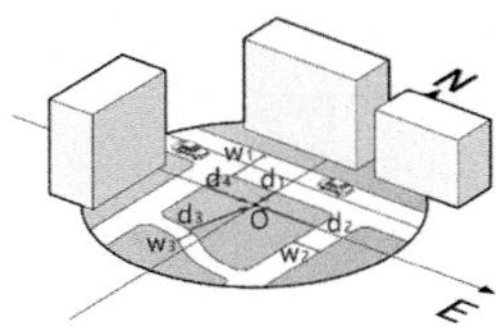	VTD = Sum ((1/distance) * length) Function of distance to a road and the road length: the effect of vehicle traffic heat emission (Quah and Roth 2012)
Local (neighborhood) scale	Categorical variables representing LCZs		
	LCZ1	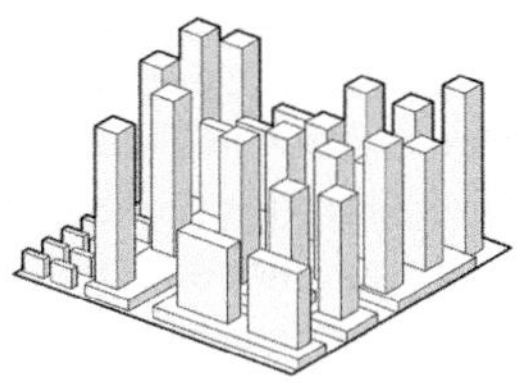	Density: very high FAR: high to very high Green space ratio: very low Land use: commercial and business, residential–commercial mixed Typology: super-tall residential towers, office skyscrapers
	LCZ2	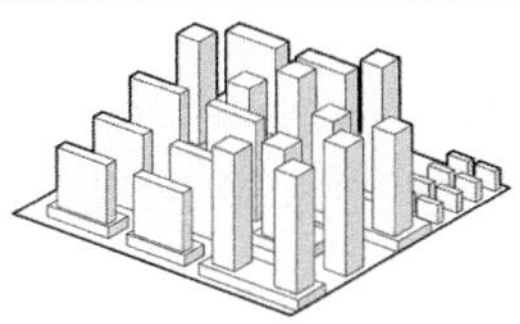	Density: high FAR: high Green space ratio: very low Land use: commercial and business, residential Typology: tall residential towers, office skyscrapers, low-rise, mid-rise housing
	LCZ3	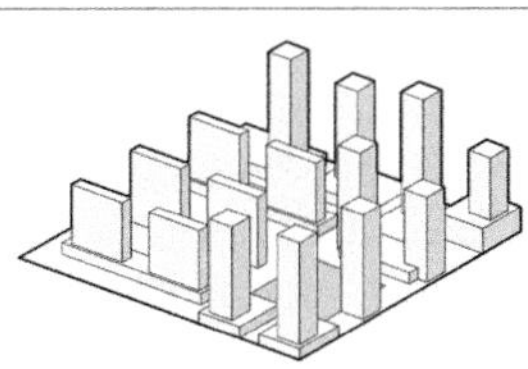	Density: medium to high FAR: medium to high Green space ratio: medium to low Land use: residential Typology: tall residential towers and slabs, neighborhood parks and open space
	LCZ4	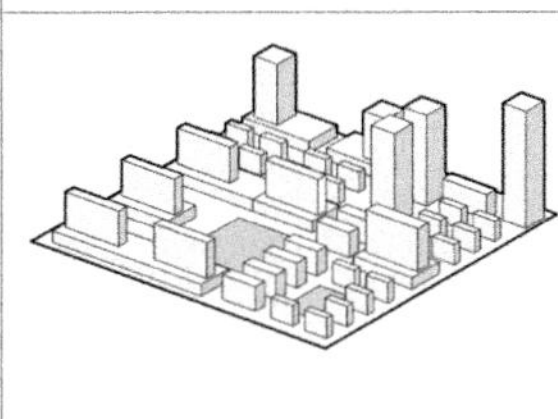	Density: medium FAR: medium Green space ratio: medium Land use: residential Typology: tall residential slabs, neighborhood parks, and open space

Note Diagrams are drafted by Feng Yipeng

Vertical porosity can be quantified using the parameter **frontal area density (FAD)**, i.e., the ratio of overlaid building façade area that is perpetual to the incoming wind direction to the site area (Chen and Ng 2011). Prescriptive design guidelines that are formulated based on rigorous scientific research are readily applicable to design scheme assessment. For instance, in Hong Kong, an indicator called the porosity ratio is manipulated to regulate the permeability (*P*) of buildings and improve ventilation. It stipulates ratios of areas of intervening and permeable elements at various height levels, as compared to the whole projected area of the assessment zone (Building Department Hong Kong SAR 2008). The Chinese Design Standard for Thermal Environment of Urban Residential Areas (JGJ286-2013) introduces an index called frontal area ratio, defined as the ratio of the frontal area of a building facing seasonal prevailing wind direction in relation to its largest possible frontal area (MOHURD (Ministry of Housing and Urban-Rural Development) 2014). Different from FAD as discussed previously, this ratio relates the building geometry and orientation with (summer) prevailing wind direction. It is required by the standard that for new residential developments in the hot summer and cold winter (HSCW) building climate zone (in which Shanghai is located), site-average frontal area ratio should be lower than 70%. Given its directionality, the index may not be very useful for square or circular building shape, but is particularly suitable for assessing site ventilation performance of linear housing form and layout: The upper limit requires that major façades be at a certain angle to the prevailing winds so as to facilitate site ventilation without compromising building ventilation.

Building height variation and street geometry (aspect ratio and length) have synthetic effects on pedestrian-level air ventilation and pollutant dispersion: Building height variation (standard deviation) causes the most change, and to less extents street aspect ratio and street length, due to the effects on horizontal and vertical mean airflow and turbulent diffusion (Hang et al. 2012). In cold climates, tall building cluster should restrict drastic changes in height between adjacent buildings (Bosselmann et al. 1995), whereas in warm/hot climates, height variance should be manipulated with enhancing pedestrian airflow and ventilation in mind.

In high-rise high-density cities, there is not much room for spacious green space and sizable parks. Small patches of green space and street trees and pocket gardens are valuable ecological assets. Vegetation provides thermal regulation through shading and evapotranspiration. The most important factor is foliage density and geometry. **Tree view factor (TVF)**, **green plot ratio (GPR)**, and **green cover ratio (GCR)** are developed to quantify the cooling effect of urban vegetation.

Other parameters include: **Hard pavement ratio (HPR)** to quantify the percentage of impervious ground surfaces, **ground surface albedo (GSA)** to estimate solar absorption and reflectance of ground substrate, **Vehicle traffic density (VTD)** for estimating the effect of anthropogenic heat emitted from road cars, and **Proximity to heat sink (DHS)** for the thermal effect of nearby sizable green space or water bodies (Table 3.3).

3.2.3 Microscale (Pedestrian-Level) Variables

- *Sky View Factor (SVF)*

The sky view factor (SVF) is defined as the fraction of the overlying hemisphere occupied by the sky. It has been used as an indicator of urban geometry and density, and their influences upon longwave and shortwave irradiation since Oke summarized negative correlation between the maximum UHI and SVF (Oke 1981). The relationship is expressed in the following equation. The sky view factor is taken in the middle of a canyon floor:

$$\Delta T_{u-r\,(\max)} = 15.27 - 13.88 \times \text{SVF} \tag{3.1}$$

When the maximum UHI occurs after sunset, this correlation reveals that UHI is significantly related to the reduced longwave radiation heat loss by the restricted view of sky (Karatasou et al. 2006). However, the maximum UHI can also exist during the day (Santamouris 2007). Uncertainty arises when attempting to apply this equation on this occasion. In high-rise Asian cities under (sub) tropical climate, due to high cloud cover, moisture, and other air particles, the diffuse radiation is a significant component of global radiation and the sky is a radiation heat source. SVF can indicate the degree of exposure to solar heat gain in daytime. So it is necessary to investigated daytime UHI and nighttime UHI separately.

The measurement methods of SVF include geometrical calculation (Johnson and Watson 1984), on-site hemispherical photograph (Steyn et al. 1986) and numerical modeling. Ratti and Richens developed a two-dimensional raster image analytical tool, by which SVF can be calculated and plotted in an urban map, using digital elevation model DEM (Ratti and Richens 2004). Urban SVF computation and visualization can be carried out on a GIS platform, e.g., (Chen et al. 2012). We further developed this method by considering the effect of vegetation on SVF. Two additional DEMs were used, which are the vegetation height, and the trunk height. In reality, it is very rare that a tree behind a tall building can still obstruct the sky; therefore, this extreme case was neglected in this study. The spatial relation given by (Gál and Unger 2014) was adopted, but using DEM for fast processing. Figure 3.4 gives an illustration of different objects that obstruct the sky along a certain direction from a given point, including tree ($T1$) with the lowest trunk elevation angle ($\angle 1$); tree ($T2$) with the highest elevation angle ($\angle 2$), its trunk elevation angle ($\angle 3$); and building (B) with the highest elevation angle ($\angle 4$) (Gal 2014).

Base on the above spatial configuration, SVF for the given point can be given as:

$$\text{SVF} = 1 - \psi_{(B)} - \psi_{(T1)} \cdot (1 - \Psi) - \psi_{(T2)} \cdot (1 - \Psi) \tag{3.2}$$

where $\psi_{(B)}$, $\psi_{(T1)}$, and $\psi_{(T2)}$ are the view factors of B, $T1$, and $T2$, respectively. Ψ is the transmissivity for trees, indicating the transparency of tree crown. In practice, the value of 0.03 was selected for shortwave radiation (Lindberg and Grimmond 2011). Following the raster-based method (Chen et al. 2012), the calculation for $\psi_{(B)}$, $\psi_{(T1)}$,

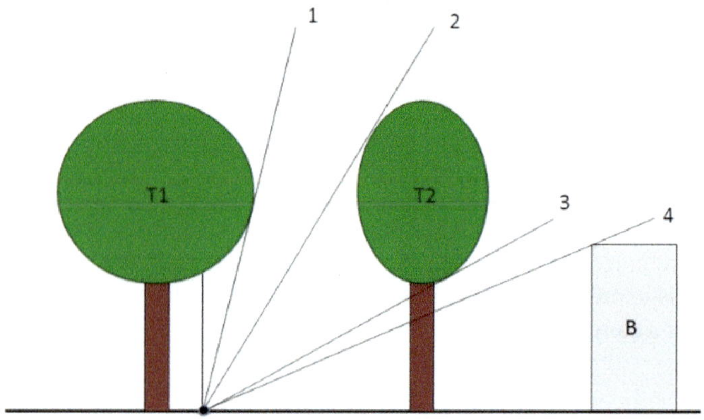

Fig. 3.4 Elevation angle for different objects that obstruct the sky along a certain direction

and $\psi_{(T2)}$ individually was fairly straightforward. For a rotation angle of a (indicating a given direction):

$$\psi_{(B)} = \sin^2(\angle 4) \cdot a/360 \tag{3.3}$$

$$\psi_{(T1)} = \sin^2(90°) \cdot a/360 - \sin^2(\angle 1) \cdot a/360 \tag{3.4}$$

$$\psi_{(T2)} = \sin^2(\angle 2) \cdot a/360 - \sin^2(\angle 3) \cdot a/360 \tag{3.5}$$

And different bigger/smaller relationship between the four elevation angles needs to be checked to ensure there is no re-calculating of obstructed sky, e.g., if $\angle 4$ is bigger than $\angle 2$, then there will be no additional obstruction to the sky caused by T2. Different elevation angle scenarios were analyzed, and the raster-based method can process the comparison fairly fast. A SVF map produced using this method is shown in Fig. 8.6a. The performance of the developed model was quite good: On a daily used PC with 3.20 GHz CPU, 4 GB memory, the model calculates the SVF of the study area (1 km by 1 km with 2 m resolution) within 1 h.

The simulated SVF values at measurement points are correlated fairly well with values calculated based on hemispherical sky view images, as shown in Fig. 6 in Yang and Chen (2016). The discrepancy could be attributed to the fact that the photographic method does not consider the transmittance of tree canopies; i.e., they are considered zero of transmissivity, which will cause underestimation of SVF.

- *Total Site Factor (TSF)*

Solar access to the ground level in a high-rise environment is subject to the interplay between solar geometry (seasonal and geographical variations in sun paths and solar intensity) and canopy geometry (building, tree, and other canopy configuration). This is particularly so on clear days when direct radiation dominates the global radiation.

It is difficult to use SVF to measure this interplay, because SVF may not be able to represent solar availability very well due to its ignorance of sun paths and reflected and diffuse radiation (Robinson 2006). For this reason, TSF is introduced to quantify the solar access at a measurement point in an integrative manner.

TSF is a function of variations in the geometry of surrounding canopies (buildings and/or trees), sun track, solar radiation intensity, and time. The TSF is calculated using Eq. (3.6):

$$\mathrm{TSF} = (\mathrm{SR}_{\text{dir-under}} + \mathrm{SR}_{\text{diff-under}}) \big/ (\mathrm{SR}_{\text{dir-over}} + \mathrm{SR}_{\text{diff-over}}) \tag{3.6}$$

where

$\mathrm{SR}_{\text{dir-under}}$ is direct radiation received under canopy;
$\mathrm{SR}_{\text{diff-under}}$ is diffuse radiation received under canopy;
$\mathrm{SR}_{\text{dir-over}}$ is direct radiation received over canopy; and
$\mathrm{SR}_{\text{diff-over}}$ is diffuse radiation received over canopy.

TSF can be calculated using WinSCANOPY software based on hemispherical sky photographs acquired by digital camera and fish-eye lens. WinSCANOPY is a digital image analyzer for canopy and solar radiations analysis (Regent Instruments Inc. 2004). Calculating TSF follows two steps: (1) canopy processing, to classify pixels into sky and canopy; and (2) canopy analysis, which applies certain theories and algorithms to analyze canopy structure and its impact on radiation. The direct radiation above canopy is a function of latitude/longitude and defined time period. The direct radiation below canopy is calculated based on the location of sun path and the pixel definitions (sky or canopy) of pixels under the sun path: It is assumed equal to the radiation above the canopy if the pixel is defined as "sky"; or assumed zero if the pixel is defined as "canopy." Because the focus on TSF is the shading effect in clear sky days, diffuse radiation above canopy is therefore assumed to be 15% of direct radiation reaching the canopy as advised by (Olgyay 1963). The diffuse radiation below canopy is the diffuse radiation above canopy multiplied by the indirect site factor, which is calculated with a formula by the software depending on the diffuse light distribution model selected.

- *Tree View Factor (TVF)*

How well the shading effect of vegetation canopy offers to measurement points directly influences the microclimate of that point. Tree canopy is capable of intercepting both direct and reflected solar radiation from above and longwave radiation from below, thus having significant impacts on both daytime and nighttime UHI intensity. The extent of tree canopy shading depends on vegetation canopy area and foliage density; therefore, it is associated with vegetation stand architecture and leaf area density of the tree canopy above the pedestrian level (i.e., 1.5 m above the ground level).

Indicators of vegetation shading have been developed and used in outdoor thermal environmental studies. For instance, tree cover shading was used to estimate the percentage of vegetation above 1 m in height (Giridharan et al. 2008). A parameter named partially shaded area (PSA) was adopted to estimate the shading effect of street trees (Shashua-Bar and Hoffman 2000). Both methods are limited in a two-dimensional and static manner. The introduction of the tree view factor (TVF) adopts a new approach to quantify the tree shading effect. Essentially, it refers to the portion of the tree canopy area that is projected on a hemispherical sky view image.

TVF is calculated based on hemispherical sky view images acquired on site using a fish-eye lens camera. TVF is defined as the fraction (or percentage when multiplied by 100) of vegetation canopy in a specified region of the hemispherical canopy above the lens (i.e., the measuring point). Considering that the vegetation canopy area close to the horizon of the hemisphere is distanced from the measurement point and incapable of providing effective shading to the point, the field of view is set within 75° in the zenith angle instead of 90°, i.e., the complete hemisphere.

In calculating TVF, the vegetation canopies are separated and the percentage is calculated. The method to identify tree canopy areas is essentially the same way to obtain SVF, only except that the pixel classification process should classify vegetation pixels as the sky. The color identification method is used to separate tree canopy areas from other areas.

- *Green Cover Ratio (GCR)*

Green cover ratio (GCR) is different from GR. GR means the greenery cover percentage of a site area, while GCR means the green cover percentage within the influential area centered by a measuring point. Therefore, GCR is intended to indicate the quantity of vegetation that immediately influences individual pedestrians.

It should be noted that on calculating GCR, besides turf and shrub areas, the vertical projection of tree canopy area on the ground surface within the specified circle area is also counted as greenery covered area. GCR is calculated using the equation as follows (Eq. 3.7):

$$\mathrm{GCR} = \sum_i G_i \bigg/ \pi R^2 \tag{3.7}$$

where G_i are the sizes of green patches (including the projected area of tree canopy) that are within the influential area centered by a measurement point; R is the radius of the circle centered by the measurement point. The circle indicates the influential area on the point. R is 20 m in the present study.

- *Green Plot Ratio (GPR)*

The green plot ratio (GPR) is derived from the leaf area index (LAI) to describe the leaf area density in a certain area instead of a single plant. LAI is defined as the total single-side leaf area of a plant species per unit of ground area. It can also be considered as the area ratio of leaves and ground cover area. The concept of GPR is proposed by Ong, which is defined as the ratio of the total leaf area of the planted landscape to the plot or site area (Ong 2003). GPR indicates the density of leaf quantity on a site. It is thus an important indicator of greenery cooling effect, because its capacities of providing shading and evaporative cooling are both directly related to the amount of green mass, i.e., the leaves on the plants. This ecological measure, with a strong resemblance with building plot ratio (i.e., FAR), is considered a single metric for the multifold ecological benefits of greenery, and more accurate than the conventional metric of green ratio.

Foliage density and mass essentially govern the daytime cooling performance of plants (Wong and Chen 2009). The maximum temperature difference between the adjacent open space and the foliage volume (under leave canopies) is only 1.2 °C for a sparse plant with the LAI of 1, but it reached 4.6 °C for a dense one (LAI = 5). Nevertheless, for sparse or medium dense plants, it was found that the leaf surface temperature is significantly lower than the air temperature under the canopies. This temperature difference can lead to sensible heat flux advection from the air to the leaves and help lower down the air temperature. On the other hand, the relatively lower surface temperature leads to a lower radiant temperature and thus improves the thermal comfort level. For the greenery on building walls, the LAI values are even more important than on the ground, because a stable air temperature under the leaves can only be achieved with a very high LAI, e.g., 5, as indicated in the experiment. LAI can be derived from another leaf metric leaf area density (LAD). It is the total single-side leaf area (in m^2) per unit of canopy volume (in m^3).

To calculate GPR, LAI values of all types of greenery (including tree, shrub, grass) and the area portions of all types of greenery are needed. It is possible to estimate the LAI of trees from a hemispherical image, but the model assumes that the object under study is a continuous infinite forest and thus does not work for a single tree or several trees in a high-rise environment. Referring to the generalized LAI values as summarized in other studies is thus considered as an alternative. In this study, the GPR of the source area of a measurement point (i.e., within 20 m radius centered by the measuring point) is calculated by the following steps: firstly, the areas of grass, shrub, middle-dense canopy trees and dense canopy trees are estimated and recorded, respectively, in site survey; secondly, values of GPR are calculated with reference to the LAI data given in (Scurlock et al. 2001; Ong, 2003).

On the GIS platform, GPR is calculated based on the definition given in (Ong 2003), which is the average leaf area index (LAI) value on a surface. The typical LAI values proposed by (Ong 2003) were selected, i.e., 1.0 for grass, 2.0 for shrub, and 5.0 for trees. Standard GIS functions were used based on DEM data for the calculation.

In addition to the above vegetation parameters, the evapotranspiration rate of vegetated areas is also influenced by soil moisture, plant type (e.g., latifoliate or coniferous), stages of plant development (e.g., young or mature), and weather. These factors are not included in the vegetation factor because of the following reasons: (1) The first three parameters are site-specific parameters. In other words, the values of these parameters are similar for all the points on the same site. Therefore, these parameters are differentiated and their impacts are compared on a site-to-site basis; and (2) the weather condition is the same for all the measurement points of the same group which are measured simultaneously.

- *Ground Surface Albedo (GSA)*

The optical characteristics of construction materials, especially the albedo to solar radiation, could have very important impacts on urban energy balance (Oke 1987). Albedo is defined as the hemispherically and wavelength-integrated reflectivity of a surface (Taha 1997b). An average urban albedo of 0.15 is estimated for mid-latitude cities in the absence of snow (Oke 1987). The effective albedo in urban areas is considerably reduced as a result of two mechanisms (Santamouris 2001): Darker buildings and urban surfaces absorb solar radiation, and multiple reflections inside urban canyons significantly reduce the effective albedo. Materials with higher albedos absorb less solar radiation and as a result keep their surfaces cooler; consequently, the intensities of longwave radiation are reduced and the surrounding and downwind area would enjoy lower air temperatures due to smaller convective heat fluxes from the cooler surfaces (Taha 1997b). Therefore, urban albedo has a significant impact on UHI formation. A study shows that an average reduction of 2 °C and up to 4 °C in air temperature, and a 10% reduction in building cooling loads can be achieved by an increase of 0.13 in the urban albedo (Taha 1997a).

The emissivity of materials is an important optical property. Materials with high emissivities emit the absorbed heat more easily by longwave radiation and cool themselves faster. However, urban materials do not vary markedly in emissivity (Santamouris 2001; Oke 1987). Furthermore, studies have shown that the role of emissivity in contributing to UHI is minor e.g., (Oke et al. 1991). Considering these, the emissivity is not used as a variable in the study.

3.2.4 Block-Scale (Site Level) Variables

- *Floor Area Ratio (FAR)*

The floor area ratio (FAR), or plot ratio used in an equivalent meaning, is the total building floor area divided by the site area. FAR is an important indicator in urban zoning and planning, to control the intensity of construction in a certain urban area. It describes the population and building density of a specific site instead of a measurement point. Therefore, FAR is adopted as the complementary indicator of building density, anthropogenic heat, and site ventilation potential.

- *Building Cover Ratio (BCR)*

Building cover ratio or building footprint has a strong impact on the roughness sub-layer (RSL) turbulence production and thus street ventilation and pollutant dispersion. Increasing BCR in general leads to smaller building spacing and thus higher street aspect ratio and directs the isolated street canyon airflow toward wake interference and skimming flow patterns.

- *Green Ratio (GR)*

The importance of GR as a variable describing the environmental effect of vegetation is self-evident, because it controls the size of a green area, thus the quantity of vegetation in a site. Increasing the size of a green area can produce the greatest reduction in ambient air temperature compared to other cooling strategies (Dimoudi and Nikolopoulou 2003). A small green area of about 60 m × 40 m shows a clear cooling effect (Saito et al. 1990/1990). A urban park with a size of 0.6 km^2 could reduce the air temperature at noon by up to 1.5 °C in a busy commercial area 1 km downwind (Ca and Asaeda 1998).

It is stipulated by the Chinese housing design code that newly developed residential quarters must have a GR of no less than 30%; and redevelopment sites for residential use must have a GR of no less than 25% (MOHURD (Ministry of Housing and Urban-Rural Development) 2002). In a recently released eco-housing rating method in China (聂梅生 et al. 2003), the GR is required to reach 35% as a prerequisite in order to get access to the eco-housing rating list.

- *Pavement Cover Ratio (PCR)*

Paved and impervious urban surface generally absorbs more solar heat during daytime and consequently releases more at night as compared to the vegetated surface, which contributes to UHI formation. Consequently, the ratio of paved area is an effective indicator for describing the imperviousness of land cover. Thermal properties of materials that are influential on UHI also include thermal conductivity, heat capacity and thermal diffusivity, which are obtained by dividing the thermal conductivity of a material by its heat capacity (Gartland 2008). A higher diffusivity leads to an enhanced heat storage during the day and gradually releases it at night, thereby contributing significantly to the nighttime UHI. The hard pavement generally has a higher thermal diffusivity than the natural (soft) land cover such as soil, vegetation, and waters (Oke 1987). It is therefore important to estimate the hard pavement ratio within the influential range of measurement points.

The use of 50-m-radius areal average is adopted as suggested by (Jusuf and Wong 2009; Kruger and Givoni 2007). Pavement cover ratio is also calculated with standard GIS functions. Land-use dataset was used to determine if a pixel is paved or not.

- *Frontal Area Density (FAD)*

The effect of building geometry on pedestrian-level wind velocity modification needs to be parameterized. Variables used to estimate surface roughness length (Z_0) include BCR (Grimmond and Oke 1998) and FAD (Chen and Ng 2011). For the urban form of dense high-rise buildings with heterogeneous heights, FAD is an important indicator of urban block porosity in the supplement to BCR, as it can provide three-dimensional information of building geometry and take into account prevailing wind direction.

The GIS-based tool developed in (Burian 2002) was used to calculate the FAD map with consideration of the wind direction. For a given planar surface, the FAD is given by:

$$\mathrm{FAD}(\theta) = A(\theta)_{\mathrm{proj}(\Delta z)} \big/ (A_T \cdot \Delta z) \tag{3.8}$$

where $A(\theta)_{\mathrm{proj}(\Delta z)}$ is the area of building surfaces projected into the plane normal to the approaching wind direction for a specified height increment Δz; θ is the wind direction angle; and A_T is the total plan area of the study site. The algorithm reads in the wind rose data and derives the total $a_f(z)$ for a site by calculating the weighted summation of the $a_f(z)$ for all wind directions.

- *Vehicle Traffic Density (VTD)*

Anthropogenic heat is an important contributor to urban warming. In city center, primary sources for anthropogenic heat emission are heat from heating, ventilation, and air-conditioning (HVAC) systems and heat induced by vehicle traffic. In commercial and business center such as Lujiazui CBD, most outdoor parts of HVAC systems (air outlet, condensing tower, etc.) are installed on podium level and service-floors of main towers; therefore, the effect of the emitted heat on pedestrian level is not significant and so only heat induced by vehicle traffic is taken into account. VTD is developed for this aspect, considering road network hierarchy and distance to road centers.

A GIS-embedded extension was developed to model the heat induced by road traffic. In practice, the VTD is influenced by various factors, and related heat emission is very difficult to predict. This study takes a simplified approach by using the road network to estimate the vehicle traffic and related heat emission. The preliminary assumptions are that the length and width of the road segment will have a proportional relationship with VTD, and the distance to a road segment will have a reverse proportional relationship with VTD. In this case, VTD for a given point is defined as:

$$\mathrm{VTD} = \sum (L(i)/D(i)) \tag{3.9}$$

where $L(i)$ is the length of the ith road segment, and $D(i)$ is the distance from the given point to this road segment. The algorithm iterates through all road segments and calculates the summation of VTD.

- *Proximity to Heat Sink (PHS)*

Water bodies and green spaces are important heat sinks that can alleviate the adverse thermal effect of urbanization. The proximity to heat sink is dependent on the shape and size of the heat sink, the distance to it, and the angle between prevailing wind direction. For instance, in the Lujiazui CBD study (Chap. 8), the two prominent heat sinks within the Lujiazui CBD are Huangpu River and Lujiazui Central Green (LCG). Because Huangpu River is to the leeward side (northwest) of Lujiazui CBD while LCG is to the windward side; therefore, LCG is chosen as the heat sink, and PHS is calculated considering wind direction and distance to the centroid of LCG (in the artificial lake in LCG).

PHS can be calculated using standard GIS Euclidean distance function, which calculates the distance from the given point to the centroid of the water body in Lujiazui Park to be considered as heat sink.

3.2.5 *Local-Scale (Neighborhood Level) Variables*

The set of local-scale variables is to control the boundary conditions of the intended microscale urban climate study, especially for an empirical one. Lowry pointed out about 40 years ago that an urban climate is the overlaid result of background, landscape, and human (including urbanization) effects (Lowry 1977). The urban and local effects could have effects on the empirical investigation carried out at the pedestrian level, i.e., the microclimate in the urban canopy layer (UCL). Therefore, it is impossible to measure the "site-specific" microclimatic effect unless those background (seasonal, geographical and multi-scale urbanization) effects can be effectively controlled.

The local climate zone (LCZ) is a scheme to divide the urban surfaces based on the effect of urban fabric, land cover, structure and metabolism (Stewart et al. 2013). A well-round scheme of LCZ would be based on comprehensive and high-resolution data of urban form, including building density, vegetation cover, canyon aspect ratio, sky view factor, mean height of roughness elements, anthropogenic heat, and typically supported by an urban GIS platform. As urban fabric of Shanghai as well as other large cities in China appears highly mixing in block density, housing typology, and land use, we adopt a simplified classification scheme based on district-level density and green space database. A detailed explanation on the LCZ scheme for Shanghai is given in Chap. 4.

Microclimate studies within the same LCZ would be considered as having comparable boundary conditions, and thus, it is possible to capture the microscale impact of urban form by empirical investigation.

3.2.6 *Climatic Factor*

The background climate has fundamental impacts on the microscale urban climate. For instance, a nearly calm, cloudless, and anticyclonic weather condition is favorable for forming the maximum UHI (Oke 1987). Increases in wind speed tend to flatten the differences caused by urban modification on microclimate, due to the turbulent, convective, and advective heat transfer processes. Cloud cover is related to UHI pattern because of its impact on the longwave radiation heat loss from ground surfaces. The influence of cloud cover on incident solar radiation is also important when considering the daytime UHI. Solar radiation is the source of energy for all the microclimate process, and its intensity determines the extent of urban thermal anomalies caused by urban material and geometry. In this light, the three parameters, i.e., solar radiation on-site (SRS), wind velocity at day and at night (WVD and WVN), and cloud cover at day and at night (CCD and CCN) are adopted as the independent variables representing the climatic influence.

References

Bosselmann P, Arens E, Dunker K, Wright R (1995) Urban form and climate: case Study Toronto. J Am Plan Assoc 61:226–239
Building Department Hong Kong SAR (2008) Sustainable building design guidelines. APP-152
Burian S (2002) Urban informatics and morphological analysis
Ca VT, Asaeda T (1998) Reductions in air conditioning energy caused by a nearby park. Energy Build 29:83–92
Chen L, Ng E (2011) Quantitative urban climate mapping based on a geographical database: a simulation approach using Hong Kong as a case study. Int J Appl Earth Obs Geoinf 13:586–594
Chen L, Ng E, An X, Ren C, Lee M, Wang U, He Z (2012) Sky view factor analysis of street canyons and its implications for daytime intra-urban air temperature differentials in high-rise, high-density urban areas of Hong Kong: a GIS-based simulation approach. Int J Climatol 32:121–136
Corbusier L (1987) The city of to-morrow and its planning. Dover
Dimoudi A, Nikolopoulou M (2003) Vegetation in the urban environment: microclimatic analysis and benefits. Energy Build 35:69–76
Eliasson I (2000) The use of climate knowledge in urban planning. Landscape Urban Plan 48:31–44
Erell E (2008) The application of urban climate research in the design of cities. In: Santamouris M (ed) Advances in building energy research, vol 2. Earthscan, London, pp 95–121
Fesselmeyer E, Seah KYS (2018) The effect of localized density on housing prices in Singapore. Reg Sci Urban Econ 68:304–315
Gal CV (2014) The influence of built form and vegetation on the canopy layer microclimate within urban blocks. Acta Climatologica ez Chorologica Universitatis Szegediensis, 47–48
Gál T, Unger J (2014) A new software tool for SVF calculations using building and tree-crown databases. Urban Climate 10(Part 3):594–606
Gartland L (2008) Heat islands: understanding and mitigating heat in urban areas. Earthscan, London
Giridharan R, Lau SSY, Ganesan S, Givoni B (2008) Lowering the outdoor temperature in high-rise high-density residential developments of coastal Hong Kong: the vegetation influence. Build Environ 43:1583–1595
Givoni B (1998) Climate considerations in building and urban design. Van Nostrand Reinhold, New York
Grimmond CSB, Oke T (1998) Aerodynamic properties of urban areas derived from analysis of surface form. J Appl Meteorol 38:1262–1292
Fung A, Lau S, Lin P, Zhang Z, Yim S, Guoxing W & 冯宜萱, 刘., 林萍英,张智栋,严汝洲,王国兴 (2011) Green and people-oriented design approach for a hyper density sustainable community: redevelopment of Upper-Ngau Tau Kok Estate phases 2&3 in Hong Kong. 以住户为导向的高层高密度可持续社区设计——香港牛头角上邨(二、三期)重建项目(in Chinese). J Eco-city Green Build, 20–27
Hang J, Li Y, Sandberg M, Buccolieri R, Di Sabatino S (2012) The influence of building height variability on pollutant dispersion and pedestrian ventilation in idealized high-rise urban areas. Build Environ 56:346–360
Johnson GT, Watson ID (1984) The determination of view-factors in urban canyon. J Climate Appl Meteorol 23:329–335

Juan Y-H, Yang A-S, Wen C-Y, Lee Y-T, Wang P-C (2017) Optimization procedures for enhancement of city breathability using arcade design in a realistic high-rise urban area. Build Environ 121:247–261

Jusuf SK, Wong NH (2009) Development of empirical models for an estate level air temperature prediction in Singapore. In: Proceeding of second international conference on countermeasures to urban heat islands, Berkley, California, USA

Karatasou S, Santamouris M, Geros V (2006) Urban building climatology. In: Santamouris M (ed) Environmental design of urban buildings: an integrated approach. Earthscan, London, U.K., pp 95–119

Kruger E, Givoni B (2007) Outdoor measurements and temperature comparisons of seven monitoring stations: preliminary studies in Curitiba, Brazil. Build Environ 42:1685–1698

Kubota T, Miura M, Tominaga Y, Mochida A (2008) Wind tunnel tests on the relationship between building density and pedestrian-level wind velocity: development of guidelines for realizing acceptable wind environment in residential neighborhoods. Build Environ 43:1699–1708

Lindberg F, Grimmond CSB (2011) The influence of vegetation and building morphology on shadow patterns and mean radiant temperatures in urban areas: model development and evaluation. Theoret Appl Climatol 105:311–323

Lowry WP (1977) Empirical estimation of urban effects on climate: A problem analysis. J Appl Meteorol 16:129–135

Mills G, Cleugh H, Emmanuel R, Endlicher W, Erell E, McGranahan G, Ng E, Nickson A, Rosenthal J, Steemer K (2010) Climate information for improved planning and management of mega cities (needs perspective). Procedia Environ Sci 1:228–246

MOHURD (Ministry of Housing and Urban-Rural Development) (2002) Standard for classification of urban green space. CJJ/T 85-2002. China Architecture and Building Press, Beijing

MOHURD (Ministry of Housing and Urban-Rural Development) (2014) Design standard for thermal environment of urban residential areas. JGJ286-2013. China Architecture and Building Press, Beijing

Ng E (2008) Policies and technical guidelines for urban planning of high-density cities—air ventilation assessment (AVA) of Hong Kong. Build Environ 44:1478–1488

Ng CY (2010) Microclimate in public housing: an environmental approach to community development. In: Ng E (ed) Designing high-density cities for social and environmental sustainability. Earthscan, London; Sterling, VA, pp 309–320

Ng E (2012) Towards planning and practical understanding of the need for meteorological and climatic information in the design of high-density cities: a case-based study of Hong Kong. Int J Climatol 32:582–598

Oke TR (1981) Canyon geometry and the nocturnal urban heat island: comparison of scale model and field observations. Int J Climatol 1:237–254

Oke TR (1987) The boundary layer climates, 2nd edn. Methuen, London and New York

Oke TR (1988) Street design and urban canopy layer climate. Energy Build 11:103–113

Oke TR, Johnson GT, Steyn DG, Watson ID (1991) Simulation of surface urban heat islands under "ideal" conditions at night Part 2: diagnosis of causation. Bound-Layer Meteorol 56:339–358

Olgyay V (1963) Design with climate: bioclimatic approach to architectural regionalism. Princeton, NJ, Princeton University Press

Ong BL (2003) Green plot ratio: an ecological measure for architecture and urban planning. Landscape Urban Plan 63:197–211

Planning Department Hong Kong SAR (2009) Urban climatic map and standards for wind environment—feasibility study—final report. Department of Architecture CUHK, Hong Kong

Quah AKL, Roth M (2012) Diurnal and weekly variation of anthropogenic heat emissions in a tropical city, Singapore. Atmos Environ 46:92–103

Ratti C, Richens P (2004) Raster analysis of urban form. Environ Plan 31:297–309

Regent Instruments Inc. (2004) WinSCANOPY 2004a for hemispherical image analysis

Robinson D (2006) Urban morphology and indicators of radiation availability. Sol Energy 80:1643–1648

Saito I, Ishihara O, Katayawa T (1990/1991) Study of the effect of green areas on the thermal environment in an urban area. Energy Build 15–16:493–498

Santamouris M (2001) Energy and climate in the urban built environment. James & James, London, UK

Santamouris M (2007) Heat island research in Europe: the state of the art. In: Santamouris M (ed) Advances in building energy research, vol 1. Earthscan, London, pp 123–150

Scurlock JMO, Asner GP, Gower ST (2001) Global leaf area index from field measurements, 1932–2000. Data set. Available on-line [http://www.daac.ornl.gov] from Oak Ridge National Laboratory Distributed Active Archive Center, Oak Ridge, Tennessee, U.S.A.

Shanghai Bureau of Planning and Land Resources (2016) Technical guidelines of control detailed planning of Shanghai 2016 edition (上海市控制性详细规划技术准则2016修订版)

Shanghai Government (2003) Technical regulations on urban planning and management of shanghai (Land use and building administration) (上海市城市规划管理技术规定(土地使用/建筑管理)

Shashua-Bar L, Hoffman M (2000) Vegetation as a climatic component in the design of an urban street: an empirical model for predicting the cooling effect of urban green areas with trees. Energy Build 31:221–235

Steemers K, Steane MA (eds) (2004) Environmental diversity in architecture. Spon Press, New York

Stewart ID, Oke TR, Krayenhoff ES (2013) Evaluation of the 'local climate zones' scheme using temperature observations and model simulations. Int J Climatol 33:1–19

Steyn DG, Hay JE, Watson ID, Johnson GT (1986) The determination of sky view-factors in urban environments using video imagery. J Atmos Oceanic Technol 3

Taha H (1997a) Modeling the impacts of large-scale albedo changes on ozone air quality in the South Coast Air Basin. Atmos Environ 31:1667–1676

Taha H (1997b) Urban climates and heat islands: Albedo, evapotranspiration and anthropogenic heat. Energy Build 25:99–103

Wen C-Y, Juan Y-H, Yang A-S (2017) Enhancement of city breathability with half open spaces in ideal urban street canyons. Build Environ 112:322–336

Wong NH, Chen Y (2009) Tropical urban heat islands: climate, buildings, and greenery. Taylor & Francis, New York

Xu C 许 (2016) Research on reasonable residential FAR in the new era: a case study of Shanghai new town (新时期城市适宜的住宅地块容积率研究——以上海的新城为例). Shanghai Urban Planning (上海城市规划), 110–114

Yang F, Chen L (2016) Developing a thermal atlas for climate-responsive urban design based on empirical modeling and urban morphological analysis. Energy Build 111:120–130

Yim SHL, Fung JCH, Lau AKH, Kot SC (2009) Air ventilation impacts of the wall effect resulting from the alignment of high-rise buildings. Atmos Environ 43:4982–4994

Yuan C, Ng E (2012) Building porosity for better urban ventilation in high-density cities—a computational parametric study. Build Environ 50:176–189

聂梅生, 秦佑国 & 江亿 (2003) 中国生态住宅技术评估手册, 北京, 建筑工业出版社

顾竹屹, 赵民 & 张捷 (2014) Exploring the Chinese New Town development: a review and outlook of New Town development in Shanghai (探索"新城"的中国化之路——上海市郊新城规划建设的回溯与展望). Urban Planning Forum (城市规划学刊), 28–36

Part II

Chapter 4
Urban Heat Islands in High-Rise Residential Quarters

Abstract *Study intent*: This case study investigates the effect of urban design factors on summertime urban heat island (UHI) intensity in ten high-rise residential quarters in the inner city of Shanghai during summer months. On-site design variables were developed to quantify the thermal impacts from density, building layout, and greenery. The design variables that were measured on site were correlated with the variation in UHI intensity during the day and night. Based on the resultant empirical models, design strategies can be formulated for summertime UHI mitigation so as to improve outdoor human thermal comfort and lower housing air-conditioning energy consumption. *Result and discussion*: Variations in UHI are in part due to site planning, building design, and greenery. Site characteristics in plot layout, density, and greenery have different impacts on UHI-day and UHI-night patterns. The overall daytime and nighttime UHI models explain up to 77 and 90% of UHI variation, respectively. On the site level, shading from either buildings or vegetation canopy is the most important factor influencing daytime UHI. The shading factor can distort and dilute behavior of other variables, e.g., green ratio and surface albedo. Nighttime UHI is more complicated due to the influence from anthropogenic heat and is significantly related to vegetation density and coverage. Considered diurnally, the semi-enclosed plot layout with a fairly high density and tree cover has the best outdoor thermal condition. To mitigate summertime UHI, it is recommended to manipulate site layout and building mass, to improve shading during the day while facilitating site ventilation at night, and to increase site vegetation cover through strategic tree planting. Beyond the site and up to the local/urban scales, urban ventilation has significant impact and therefore urban wind corridor would benefit UHI mitigation on the urban scale.

4.1 Introduction

This case study aims to establish statistical relationships between measured outdoor thermal anomalies (indicated by urban heat island intensity UHI) and urban design parameters in selected high-rise high-density housing quarters in the inner

The contents in this chapter are published in Yang et al. (2010, 2011).

F. Yang and L. Chen, *High-Rise Urban Form and Microclimate*,
The Urban Book Series, https://doi.org/10.1007/978-981-15-1714-3_4

city of Shanghai, China. The hypothesis is that, with the urban- to local-scale UHI "background" effect being carefully controlled, a significant part of outdoor thermal anomalies can be explained by critical on-site urban design factors.

4.1.1 The Background Effect Control

UHI pattern is dynamic with seasonal change in background climate (i.e., radiation, temperatures, air humidity, rural soil moisture, wind patterns, etc.) (Chow 1982; Yan and Xu 1996a) and spatial variation in geographic location (i.e., urban central to suburb to rural) (Kolokotroni and Giridharan 2008). To build up a sample of cases that is deemed "representative" of form, fabric and land cover of high-rise housing quarters in Shanghai, this case study included a total of ten high-rise residential quarters distributed within the inner city of Shanghai. For an empirical study, the urban and local effect could mingle with the site effect as measured at the pedestrian level. It is impossible to measure the "site-specific" microclimatic effect unless those background (seasonal, geographical, and multi-scale urbanization) effects can be effectively controlled. The study therefore used the following precautions:

- Topographically, all the sites are located within the inner city of Shanghai, and the effect of altitude is assumed marginal, as Shanghai is a flat city situated on a fluvial plain.
- At the local scale, sites were selected, grouped, and measured according to a specific local climate zone (LCZ) scheme; the sites within each LCZ are bounded with similar landscape elements (i.e., water bodies, urban parks, highways, etc.). Each group of sites was measured simultaneously.
- Seasonally, the period of field measurement is classified into two weather periods, i.e., peak summer days and late summer days, based on meteorological data of cloud cover and air temperature.

Using data available, a LCZ map of inner-city Shanghai is produced based on the following assumptions:

- Levels of urban- to local-scale climatic effect can be classified by the LCZ mapping.
- The measured climatic variation in the experimental sites within each LCZ can thus be largely explained by on-site design variables.

The inner-city Shanghai is defined by the boundary of the inner ring highway and comprises ten administrative districts: Huangpu, Jing'an, Luwan, Xuhui, Changning, Putuo, Zhabei, Hongkou, Yangpu, and Pudong New District. Except for Huangpu, Luwan, and Jing'an, the other seven districts are only partly located within the inner ring domain (Fig. 4.1). Based on the information provided by the Shanghai Statistics Bureau at the district level, a simplified local climate zoning is carried out according to two aspects, i.e., urban cover (i.e., built vs. vegetated) that is directly related to

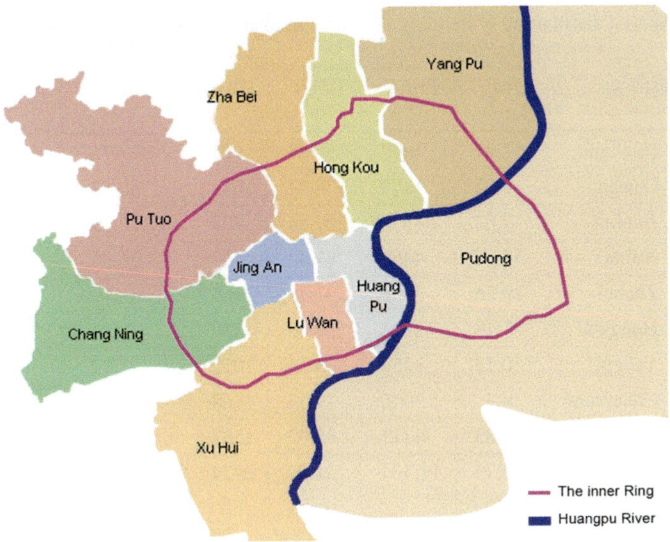

Fig. 4.1 Ten inner-city districts in Shanghai bounded by the inner ring highway

the extent of surface permeability and urban structure that is closely associated with wind, solar energy balance, and the nocturnal heat island (Oke 2004). Specifically, urban cover is quantified by **green ratio** and **residential green space density**; urban structure is quantified by **total building floor area ratio** and **high-rise building floor area ratio** (Table 4.1).

Considering the climatic parameters, there is negligible difference in solar radiation and precipitation within the inner city. The summertime wind pattern, however, could significantly affect the temperature distribution: The southeasterly winds prevailing in the summer months bring cool air from the sea and are gradually warmed up along its way into the urban area. Consequently, a temperature gradient can be formed within the Shanghai metropolis, with the temperature becoming progressively higher from southeast to northwest. This is supported by a comparison among the seven rural observatories, based on the recorded mean daily maximum temperatures during the summer period: The observatories can be divided into three groups which form three parallel belts, with the temperature gradually increasing from southeast to northwest (Table 4.2). Therefore, the districts at the northwest part can be warmer than the southeast districts. The zoning scheme should take this into consideration.

The inner city is thus divided into four LCZs:

- **LCZ 1**, the southwest (SW) zone, consists of Huangpu, Luwan, Jing'an, and the inner-city part of Xuhui. Although Xuhui district has a quite high greenery density and low built density as a whole, the density, and impervious ratio of its inner-city part, the Xujiahui area can be drastically higher than the above values. Therefore, this area of Xuhui is included in the SW zone.

Table 4.1 Spatial distribution of building floor area ratio and green space density in ten inner-city districts

Name of districts		Area (km^2)	TBFA	TFAR	HRBFA	HRFAR	GR	RGSD
Puxi (west river)	Huangpu	12.41	1922	154.88	965	77.76	2.6	0.12
	Luwan	8.05	1278	158.77	649	80.62	4.0	0.33
	Jing'an	7.62	1538	201.84	909	119.29	3.7	0.43
	Xuhui	54.76	4459	81.43	1427	26.06	13.4	1.59
	Zhabei	29.26	2684	91.73	764	36.19	7.7	0.69
	Hongkou	23.48	2992	127.43	1059	32.54	6.9	0.77
	Yangpu	60.73	4454	73.34	1052	17.32	19.4	1.01
	Changning	38.3	3046	79.53	1368	35.72	17.8	1.74
	Putuo	54.83	4711	85.92	1582	28.85	15.3	0.75
Pudong (east river)	Pudong ND	532.75	11,504	21.59	2724	5.11	25.6	2.30

Note *TBFA*—total building floor area ($\times 10^4$ m^2); *TFAR*—total building floor area ratio ($\times 10^4$ m^2/km^2); *HRBFA*—high-rise building floor area ($\times 10^4$ m^2); *HRFAR*—high-rise building floor area ratio ($\times 10^4$ m^2/km^2); *GR*—green ratio; *RGSD*—residential green space density (m^2 per capita)

Table 4.2 Matrix of mean daily maximum temperature differences (in °C) of seven rural observatories in Shanghai

Name of observatory		Jiading	Qingpu	Songjiang	Baoshan	Fengxian	Nanhui	Jinshan
	Mean	35.1	34.8	34.6	34.4	33.2	32.8	32.6
Jiading	35.1	***0***	***0.3***	0.5	0.7	1.9	2.3	2.5
Qingpu	34.8	***−0.3***	***0***	0.2	0.4	1.6	2	2.2
Songjiang	34.6	−0.5	−0.2	***0***	***0.2***	1.4	1.8	2
Baoshan	34.4	−0.7	−0.4	***−0.2***	***0***	1.2	1.6	1.8
Fengxian	33.2	−1.9	−1.6	−1.4	−1.2	***0***	***0.4***	0.6
Nanhui	32.8	−2.3	−2	−1.8	−1.6	***−0.4***	***0***	0.2
Jinshan	32.6	−2.5	−2.3	−1.9	−0.7	−0.6	−0.2	0

- **LCZ 2**, the northeast (NE) zone, comprises the inner-city part of the districts of Yangpu, Hongkou, and Zhabei.
- **LCZ 3**, the northwest (NW) zone, covers two districts Putuo and Changning.
- **LCZ 4**, the east river (ER) zone, covers the inner-city part of Pudong New District.

Clearly, the above LCZ map is a compromise due to its ignorance of variance within each zone. It is a district-level average. A thermal map produced by Ding et al. (2002) is referred to, therefore, in an effort to refine the LCZ map. The distribution of summertime mean daily maximum air temperatures was analyzed, based on measurements made by a network of 31 weather stations, in July and August for 30 days. The maximum air temperature is analyzed because of its relevance to thermal comfort and building cooling loads (Yan and Xu 1996a). The extrapolated thermal map gives an estimation of summertime peak air temperature distribution in Shanghai.

Ding's thermal map reasonably reflects the impacts from land use, density, and anthropogenic heat on air temperature distribution. For instance, the peak temperature area at Yuyuan Garden corresponds to the traditional Shanghai Lilong housing area featuring a very high building coverage and virtually absence of greenery. Two steel plants in the northern part produced markedly higher temperatures, while the Zhongshan Park at the western side created a clear temperature reduction, demonstrating an "oasis" effect.

4.1.2 Site Selection

The site selection is based on the LCZ map, while locations of the sites within each LCZ are decided with reference to the superimposed Ding's thermal map (Fig. 4.2). The typical layouts of the selected sites include linear, long linear, semi-enclosed, and interspersed (Fig. 1.4).

4.1.3 Measurement Protocol

The data collection comprises climatic measurement and site survey. Site measurement combines the use of mobile traverses and fixed weather stations. The air temperature (Ta), relative humidity (RH), wind velocity (WV), wind direction (WD), and global radiation (SRS) were measured by fixed Hobo weather stations and mobile microstations from 1 pm to 10 pm during the measurement days.

The site survey was conducted under similar weather conditions as the climatic measurement. Hemispherical sky view images were acquired using a digital camera with a fish-eye lens, and albedo values were measured using a set of Kipp & Zonen radiometers. The surface properties of the ground area within a 20-m-radius circle centered by a measurement point were elaborately recorded using a survey form.

The set of independent variables is selected to quantify the thermal influence of on-site building form, density, and greenery, as well as the background UHI effect. It includes *microscale parameters: sky view factor (SVF), total site factor (TSF), tree view factor (TVF), green plot ratio (GPR), green cover ratio (GCR), ground surface albedo (GSA), wind velocity (WVD, WVN); site-scale parameters: floor area ratio (FAR), green ratio (GR), solar radiation (SRS); and a set of categorical variables to represent the result of LCZ.* A detailed description of definitions and measurement approaches of the independent variables is given in Chap. 3 (Table 3.3).

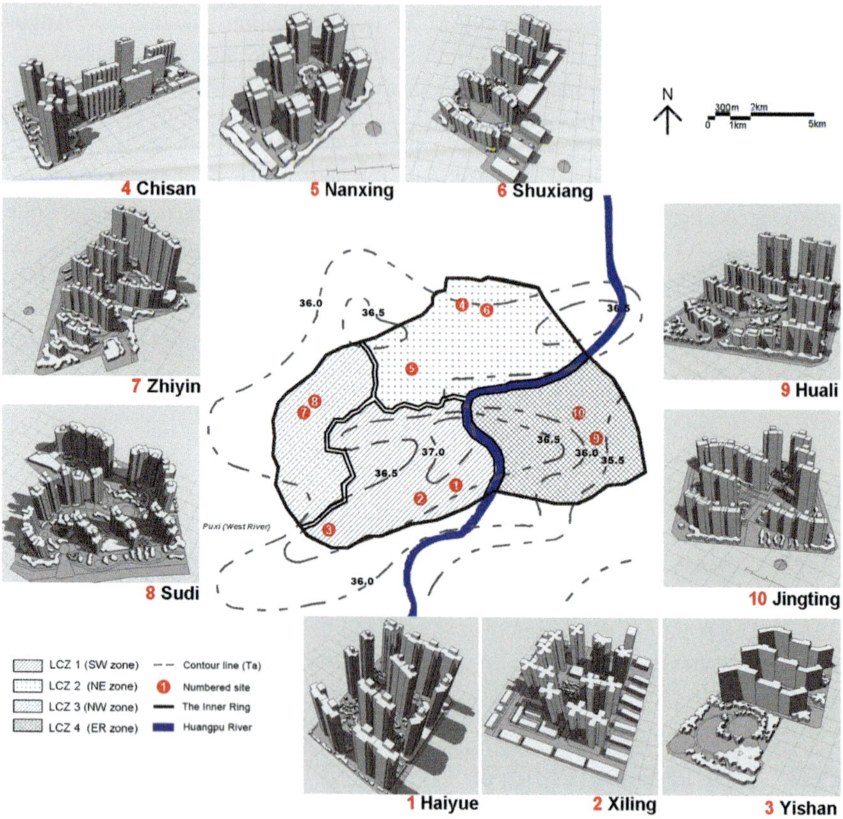

Fig. 4.2 Locations of the selected sites in the inner city based on the LCZ results

4.2 Results

4.2.1 Descriptive Statistics at the Site and Zone Levels

A cross-comparison of the site-mean UHI among different LCZs (Fig. 4.3) indicates that, concerning the impact of density and layout, the interspersed sites with fairly high densities and low green ratios exhibited lowest daytime UHI at LCZ level, but they converted to the hottest ones at night (e.g., Xiling in UCZ 1 and Nanxing in LCZ 2). The linear sites, with relatively low densities but high green ratios, showed the highest UHI-day and lowest UHI-night (e.g., Yishan in LCZ 1 and Shuxiang in LCZ 2). Note that although Yishan and Shuxiang are high in site green cover (GR), the average leaf density and tree canopy cover are in fact the lowest, indicating the dominance of unshaded open areas covered with grass and mixed with bare soil.

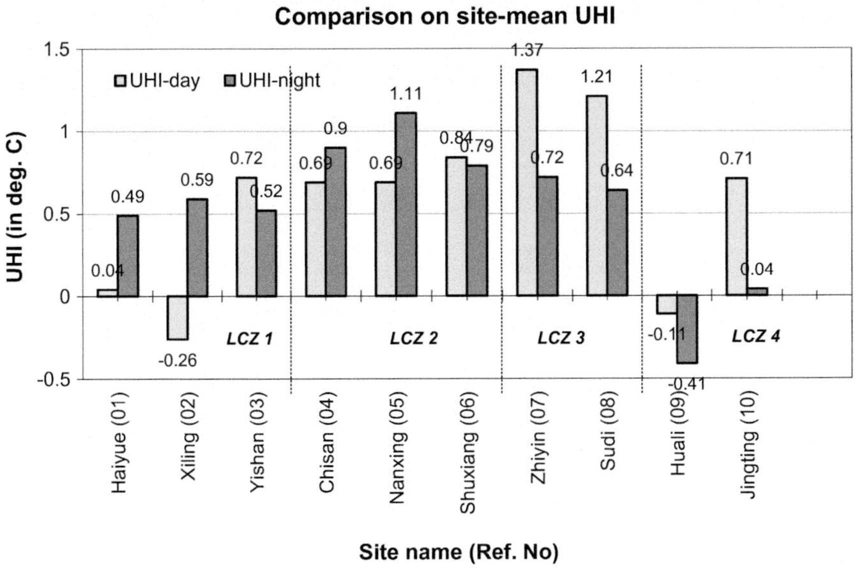

Fig. 4.3 Site-mean UHI values among ten sites

Actually, when density and layout are comparable, a site with higher percentage of dense tree cover showed markedly cooler conditions both at day and at night (Huali in LCZ 4, compared to Jingting in the same LCZ). The highest mean UHI-day of about **1.4** °C was recorded at Zhiyin in LCZ 3 and the highest UHI-night of about **1.1** °C at Nanxing in LCZ 2.

A comparison of LCZ-wise distribution of UHI intensities indicates clearly the background effect (Fig. 4.4). LCZs 1 and 2 were both measured within the peak summer day period, while both LCZs 3 and 4 were measured during the late summer day period. The differences between UCZs 1 and 2, and between UCZs 3 and 4 can be attributed to the differences in density and land cover. For instance, LCZ 3 is higher than LCZ 4 in building density while lower in green ratio. This could explain in part the overall higher UHI at LCZ 3 than LCZ 4 by about 1 °C. Furthermore, the inter-LCZ differences in UHI may be strongly influenced by the regional wind. The warmer states of LCZs 2 and 3 can be attributed to their leeward-side locations with respect to the prevailing southeast winds in summer. On the urban scale, it is possible that this wind-induced thermal gradient could override local differences in thermal properties (e.g., land use, form, and density) and heavily influence the citywide UHI pattern.

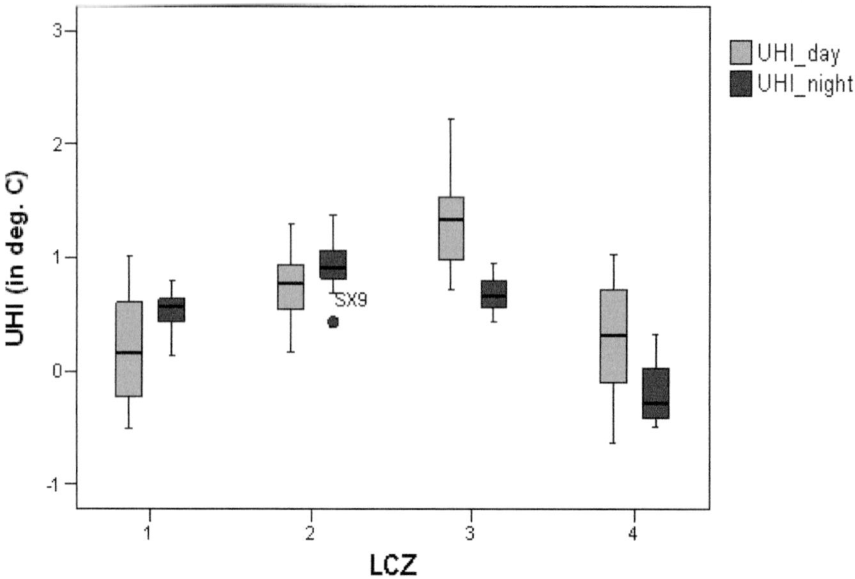

Fig. 4.4 Box plot of UHI distribution among four LCZs

4.2.2 Time Series of Temperature and Wind Velocity at the Microscale

It is helpful to gain a better understanding of any causal relationship by examining the time series development of microclimate parameters at strategically located points, before statistical analysis. In this section, the three sites in LCZ 1 are selected, the continuously measured air temperature and wind speed by fixed Hobo weather stations are plotted, and the microscale parameters related to form, density, and land cover are tabulated, for a close examination (Fig. 4.5).

Site 1—Haiyue (semi-closed) was completed in 2004. Buildings use reinforced concrete (RC) frame and are finished with light-brown ceramic tiles and green-tinted glass aluminum frame windows. Pedestrian and vehicle circulation routes do not intersect, so the open and green spaces within the estate can be used free from mobile traffic disturbance. A high degree of diversity and richness in soft- and hard-paving elements, Site 1 has large amount of grass lawns with a fairly high cover of dense shrubs and tree cover. The paving system is a mix of color and fabric, including pink and gray granite, gray and dark ceramic tiles and cement bricks. Pebble is also used at playground paths.

Site 2—Xiling (interspersed), completed in the 1980s, is part of an urban renewal program. Ten high-rise apartment towers are packed into a site of 4.8 ha. The 33–35 story buildings reach a height of nearly 100 m. The buildings are RC framed and finished with a light-colored mosaic. The affiliated facilities include a kindergarten, a market, and two estate management office buildings. Roads in the site are largely

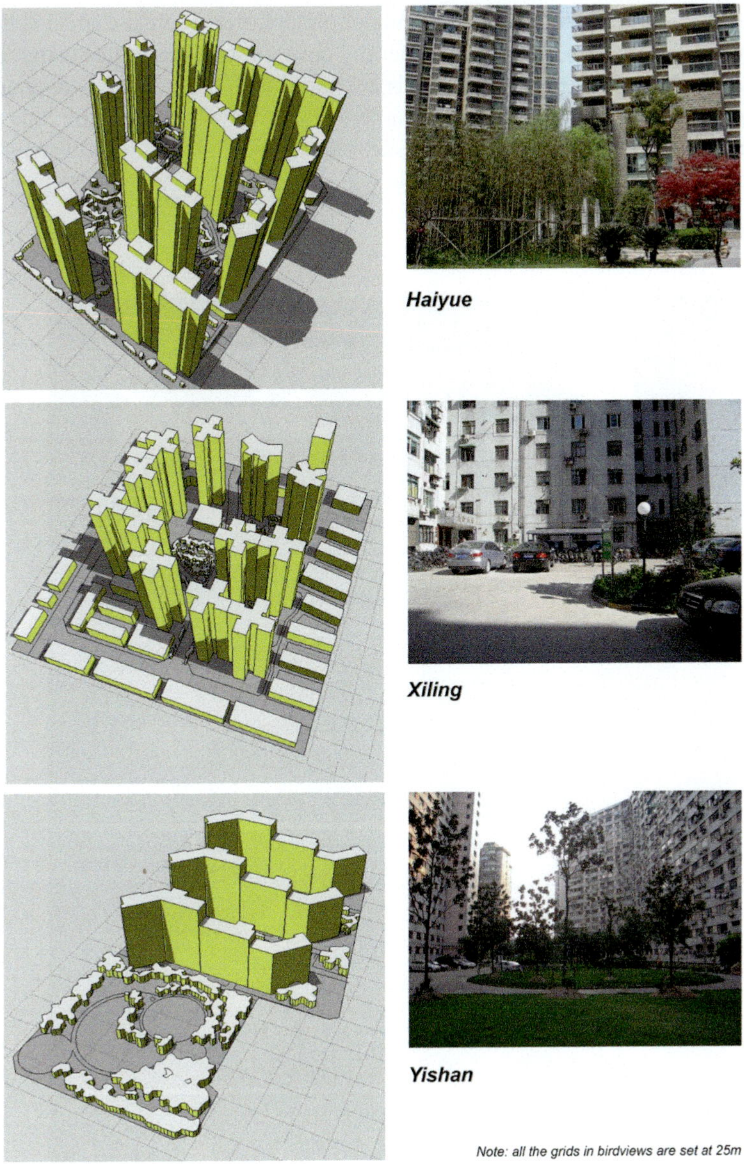

Fig. 4.5 Bird's-eye views (seen from southeast) and site photographs of the Sites 1, 2, and 3

concrete paving, while sidewalks are paved with concrete blocks or cement rendering. There are two small pieces of open spaces planted with tall southern magnolia trees. These two well-shaded areas serve as the community space of the estate and are very popular among senior residents.

Site 3—Yishan (long linear) was completed in the 1990s. It comprises three long linear, 14–18 story buildings arrayed from north to south. The buildings are walled with precast slabs and finished with white or pink plaster. There are some vegetated spaces in between these buildings, generally covered with grass and scattered with some magnolia or camphor trees. Roads are paved with light-colored concrete. The ground surface photographs and hemispherical sky view images of the four points measured with fixed Hobo stations at Site 1 are shown in Fig. 4.6 and values of six independent variables are given in Table 4.3.

- Site 1: semi-closed

Fig. 4.6 Ground surface photographs and hemispherical sky view images of P2, P4, P10, and P14 at Site 1

Table 4.3 Independent variables of P2, P4, P10, and P14 at Haiyue

Variables	SVF	TSF	GSA	TVF	GPR	GCR
P2	36.00	77.87	0.17	13.93	1.13	98
P4	35.23	88.59	0.13	7.57	1.10	80
P10	19.51	63.06	0.08	3.33	1.70	70
P14	14.19	37.84	0.05	47.62	4.05	90

Note SVF—sky view factor, *TSF*—total site factor, *GSA*—ground surface albedo, *TVF*—tree view factor, *GPR*—green plot ratio, *GCR*—green cover ratio

As is clearly seen from Fig. 4.7 that in Site 1, P14 (in light gray, bold) is the lowest in air temperature (Ta) from 1 pm to 6 pm. Ta of P2 (in black, thin) and P4 (in dark gray, bold) are very close, and both are the highest. Ta of P10 (in black, dashed) is in the middle. During the measurement period, the highest Ta reached 36.8 °C, occurred at 1 pm at P4. The maximum Ta difference is about 1.2°C, occurred at about 3:30 pm, between P2 and P14.

Comparing the Ta trend with independent variables in Table 4.2 suggests that Ta variation is positively related to SVF and TSF, while negatively related to TVF and GPR. The coolest point P14 enjoys a high degree of tree shading, while P2 and P4 are both exposed to direct sunshine when the radiation intensity is the strongest in the day.

TSF indicates the ratio of the solar radiation received under a canopy to that above the canopy. Higher TSF indicates higher amount of solar energy received at ground level. Albedos of surfaces being similar, higher TSF will thus lead to higher net radiation at ground. Referring to Oke's model (Eq. 4.1) (Oke 1987), if man-made heat, latent heat, stored heat, and advection heat transfer are same among points, sensible

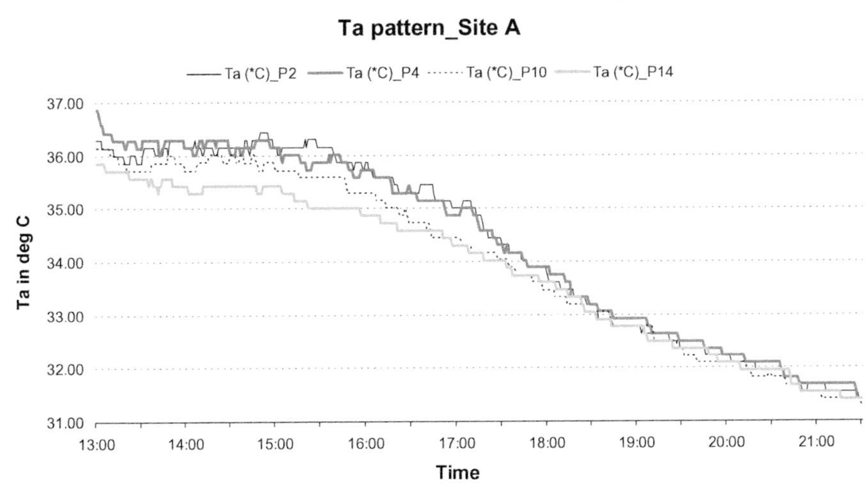

Fig. 4.7 Three-day mean temperature patterns of P2, P4, P10, and P14 at Site 1

heat fluxes would be enhanced, which leads to higher air temperature (Hoyano et al. 1999). A higher TVF and GPR tend to reduce TSF by vegetation canopy and as a result lower the air temperature.

$$Q * +Q_F = Q_H + Q_E + \Delta Q_S \mid \Delta Q_A \quad (4.1)$$

After 6 pm, Ta lines gradually mix together, although a subtle divergence can be identified between P4 and P10, where the maximum difference is about 0.5 °C. The two points are similar in GCR and TVF but differentiated in SVF and TSF. But during this period, no clear relation is found between Ta and any listed independent variable among four points.

The wind pattern is shown in Fig. 4.8. It shows the same order with Ta pattern. P2 and P10 have the highest wind velocities, while wind velocity at P14 is significantly lower than the other points. WVD variation is positively related to SVF and TSF, while negatively related to TVF and GPR.

A higher SVF or TSF indicates a higher degree of encloseness of the measurement point by built or greenery elements; consequently, air movement is more likely to be stagnated because of the wall effects of these elements. Especially, because of the local prevailing southeastern winds in summer days, site openings to southeastern direction could also facilitate natural ventilation, as shown in the cases of P2 and P10 (SVI in Fig. 4.6). The ground surface photographs and hemispherical sky view images of the four points measured with fixed Hobo stations at Site 1 are shown in Fig. 4.9 and values of six independent variables are given in Table 4.4.

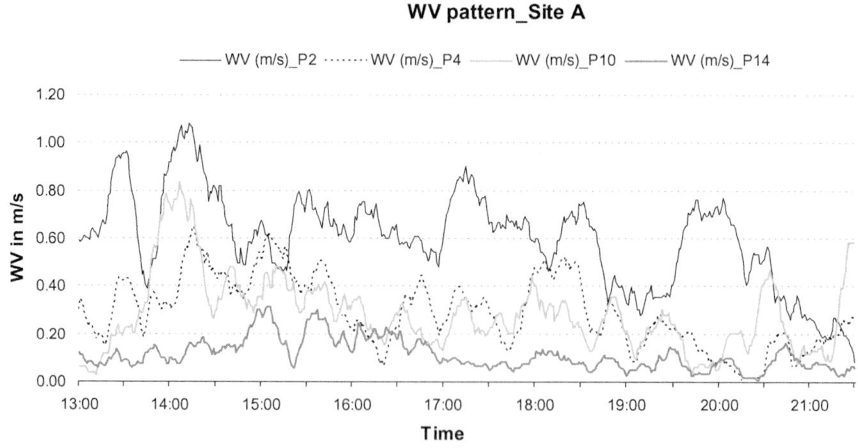

Fig. 4.8 Three-day mean wind velocity patterns of P2, P4, P10, and P14 at Site 1 (plotted as the mean values of every following 15 min)

Fig. 4.9 Ground surface photographs and hemispherical sky view images of P2, P3, and P11 at Site 2

Table 4.4 Independent variables of P2, P3, and P11 at Xiling

Variables	SVF	TSF	GSA	TVF	GPR	GCR
P2	12.91	13.34	0.14	62.42	4.30	70
P3	26.60	63.63	0.18	32.82	1.80	45
P11	5.61	6.08	0.11	36.40	4.40	70

Note SVF—sky view factor, *TSF*—total site factor, *GSA*—ground surface albedo, *TVF*—tree view factor, *GPR*—green plot ratio, *GCR*—green cover ratio

- Site 2: interspersed

Ta pattern in Fig. 4.10 shows that in the daytime, P11 has the lowest Ta while P3 has the highest Ta. The maximum difference reaches about 0.8 °C at 1:30 pm. The highest Ta at P2 and P11 occurs around 2:30 pm, one and a half hours later than that of P3. Relating to data in Table 4.4, it is found that the Ta variation is positively related to SVF, TSF, and GSA, while negatively related to GCR and GPR. As for

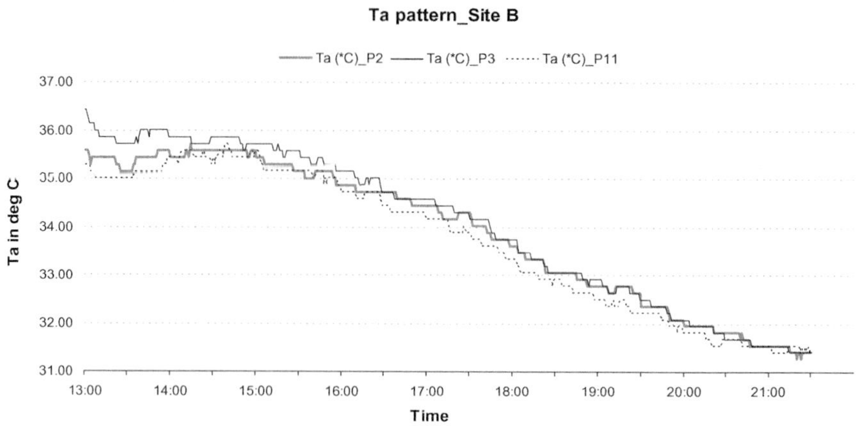

Fig. 4.10 Three-day mean temperature patterns at three points in Site 2

TVF, although TVF of P2 is much higher than that of P11, Ta at P2 is slightly higher than P11. P3 and P11 are similar in values of TVF, but P3 is significantly higher than P11 in Ta.

The impacts of SVF, TSF, and GPR on Ta are explained in Site A. For P2 and P11, other greenery variables (GCR and GPR) are similar but TVF is significantly higher in P2 than in P11; nevertheless, P11 is in some sort lower than P2 in Ta. This can be explained by the shade provided by surrounding buildings at P11. The buildings and trees work together to shade P11 and lead to a very low solar admittance (TSF = 6.08) compared with P2 (TSF = 13.34). The comparison between P2 and P11 suggests that shading by buildings is no different from shading by trees. The confusion of TVF effect on Ta between P3 and P11 can be explained by the different placements of tree canopies in the sky view images with respect to sun tracks and solar intensity variations with the sun tracks (see hemispherical sky view images of the three points in Fig. 4.9). P3 and P11 are similar in TVF, but tree canopy over P3 is largely within 45–90° in zenith angle, whereas tree canopy over P11 is well covered within the sky area of 0–45° in zenith angle, intercepting the most intensive period of direct solar radiation of a day (the red part of the sun tracks in Fig. 4.9). The different shading effects can be evidenced by the TSF values. TSF is 63.63 at P3, but it is only 6.08 at P11. The comparison between P3 and P11 indicates that TVF may not quantify shading effect very well due to interplay between solar geometry and tree canopy geometry. Other variables, e.g., TSF, may better quantify this effect.

Tree canopy shading helps delay occurrence of the maximum temperatures. Theoretically, higher albedoes should lead to lower surface and thus air temperature because of lower net shortwave radiation. The positive relationship between ground albedo and Ta here may be due to that the incident solar radiation is differentiated by building/tree canopies. Although P3 has the highest albedo, it is exposed under the highest radiation energy input. In the meantime, radiation energy input at P3 and P11 is greatly reduced by tree/building canopies. Therefore, even P3 reflects a higher portion of shortwave radiation, and the net radiation it receives is still much higher than those of P2 and P11. So *in the complex high-rise environment, a single measure of ground albedo seems not to be able to explain the Ta differences without consideration of other variables, especially the ones describing the canopy shading effect.*

After 6 pm, P11 remained the coolest point. But P2 gradually comes closer to P3 after 4:30 pm and two lines almost merge together after 6 pm. The dense tree canopy over P2 should contribute to this change because it reduces longwave heat loss to the sky. But differences in greenery cover (GCR), tree canopy (TVF), leaf density (GPR), and sky view factor (SVF) do not result in marked temperature difference between P2 and P3. Site observation found that P2 and P3 are located in the central community space of Site B, and it is closely adjacent to two major driveways that are particularly busy in the early evening. So, heat released by vehicles can contribute to the elevated temperatures at P2 and P3 and negate the behaviors of on-site variables. The fact that P2 and P3 are hotter than P11 at night can be explained by the nearby anthropogenic heat and by the less solar radiation absorbed at P11 at daytime and in result less irradiated heat at night (Fig. 4.10).

It is seen from Fig. 4.11 that WV is significantly higher at P3 than at P2 at all times (WV data at P11 is lost due to equipment disorder). The average WV at P3 is about 0.5 m/s, while it is only around 0.15 m/s at P2. The WV variation is positively related to SVF and TSF, while negatively related to TVF and GPR. It should be noted that locations of P2 and P3 are very close, only about 25–30 m, and both stations are in the center area of the Site B. As a result, the difference in the effects of buildings on wind pattern should be marginal, so the marked air stagnation at P2 should be largely caused by the surrounding tree barrier. The ground surface photographs and hemispherical sky view images of the four points measured with fixed Hobo stations at Site 1 are shown in Fig. 4.12 and values of six independent variables are given in Table 4.5.

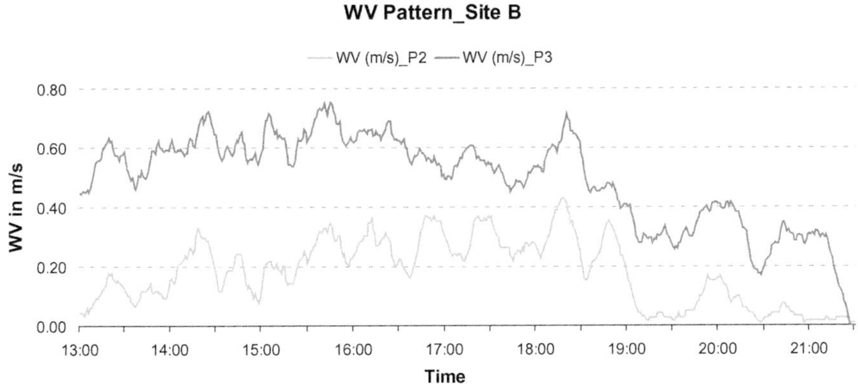

Fig. 4.11 Three-day mean wind velocity patterns of P2, P3, and P11 at Site 2 (WV of P11 is lost due to instrument disorder)

Fig. 4.12 Ground surface photographs and hemispherical sky view images of P4, P8, P11, and P12 at Site 3

- Site 3: long linear

Ta pattern at Site 3 is shown in Fig. 4.13. During daytime, P11 remains the lowest from 1 pm to 3 pm. The maximum Ta difference is about 0.8 °C. After 3 pm, Ta at all the points has a tendency of convergence. P11 has the highest TVF but the lowest SVF, TSF, and GCR. Similar to Site B, P11 reaches its maximum Ta about 2.5 h later than the other points because of the shading by trees. P12 is 100 percent covered with lawn (GCR = 100), but its Ta is significantly higher than P11, which has a GCR

Table 4.5 Independent variables of P4, P8, P11, and P12 at Yishan

Variables	SVF	TSF	GSA	TVF	GPR	GCR
P4	32.77	87.50	0.14	6.64	1.36	90
P8	30.41	84.05	0.13	15.28	1.35	90
P11	23.28	25.39	0.16	55.21	1.30	40
P12	63.21	97.97	0.14	0	1.00	100

Note SVF—sky view factor, *TSF*—total site factor, *GSA*—ground surface albedo, *TVF*—tree view factor, *GPR*—green plot ratio, *GCR*—green cover ratio

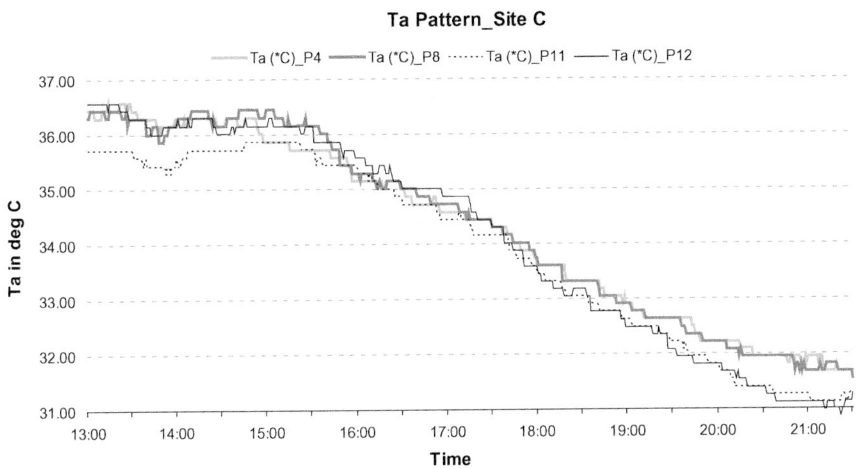

Fig. 4.13 Three-day mean temperature patterns of P4, P8, P11, and P12 at Yishan

of only 40%, so are P4 and P8. Comparison of Ta against greenery cover (GCR) among four points shows that higher ground greenery covers do not necessarily lead to lowered air temperatures.

Although SVF shows not much difference among P11, P4, and P8, TSF is much higher at P4 and P8 compared with P11. This is caused by the proper placement of tree canopy over P11 which intercepts direct sunshine considerably (as shown in SVI of P11 at Fig. 4.12). Lower TSF at P11 suggests TSF as a better shading factor than SVF in this case. Ta at P8 becomes significantly higher than P4 from 2:45 pm to 4 pm, despite that the two points share almost same values in every variable (Table 4.5).

P11 and P12 become gradually lower in Ta than P4 and P8 after 6 pm after sunset. The maximum Ta difference reaches 0.8 °C at about 8:45 pm. No clear relations can be established between this Ta gap and any independent variable. The obvious setting difference is that P11 and 12 are in fact out of the Site C sub-sector, in the park, but P4 and P8 are located within deep canyons flanked by long and bulky tall buildings. The irradiated heat from surrounding building walls and anthropogenic heat released from building indoors can contribute to the heightened temperatures.

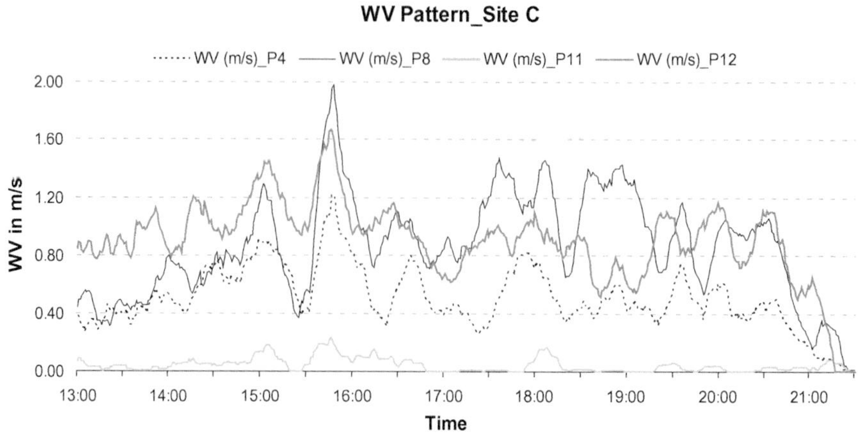

Fig. 4.14 Three-day mean wind velocity (WV) patterns of P4, P8, P11, and P12 at Site 3 (plotted as mean values of every following 15 min)

WV pattern at Site C is shown in Fig. 4.14. P11 suffers the lowest wind speed at all times, while the nearby P12 enjoys the highest during most of the daytime. The surrounding trees and dense shrub hampered the air movement at P11 (Fig. 4.12). For the other two points, it is found that P8 is higher than P4 at most time during the day. Considering the resemblance in all the variables, the divergence should be caused by the different point locations on the site: Under the prevailing southerly winds in measurement days, P4 is subjected to more serious wall effects by two linear bulky buildings compared with P8, to which the incoming winds only have one block in its way.

- Summary

The case study on Hobo station-measured points indicates the following results:

For all the three sites, comparison of Ta-day variation among the stations shows that *Ta is positively related to viewable sky* (*SVF*) and *solar admittance* (*TSF*) *while negatively related to tree canopy cover (TVF) and leaf density* (*GPR*). No coherent relations are found for ground greenery cover (GCR) and ground albedo (GSA) at all sites. The data suggests that *daytime air temperature in high-rise built environment is most influenced by the regulating coefficient on solar radiation energy input by the upper hemispherical canopy, be it natural or built. This shading coefficient is best indicated by TSF and SVF.* TVF as a shading factor is subjected to its limitation due to the interplay between tree canopy placement and solar geometry, whereas two-dimensional surface properties such as greenery coverage and albedo show no clear relations with Ta-day variation.

No clear linear relationship is identified between Ta-night variation and any independent variables. Data from Sites B and C suggests other heat sources that are not counted well by any independent variable, probably anthropogenic heat, and contributes significantly to the nighttime temperature differences.

4.2.3 Bivariate Regression at the Local Scale

The bivariate regression found that SVF, TVF, and GPR are significantly related to UHI-day, while GPR and GCR are significantly related to UHI-night in three out of the four LCZs. The impact of WVD and WVN is significant in LCZs 2 and 3, but in LCZ 1 the WVN appeared positively related to UHI-night. None of the four LCZs indicates a significant relationship of GSA with UHI-day or UHI-night. Only TSF consistently shows positive relationships with UHI-day in all the four UCZs at the 5% level. Overall, the results of UCZ-wise analysis indicate that SVF, TVF, GPR, and TSF are important in determining UHI-day, and GPR and GCR are important in determining UH-night.

Scatter plots and correlations of SVF, TSF, TVF, and GPR with UHI-day are shown in Fig. 4.15. SVF and TSF are positively related to UHI-day, while TVF and

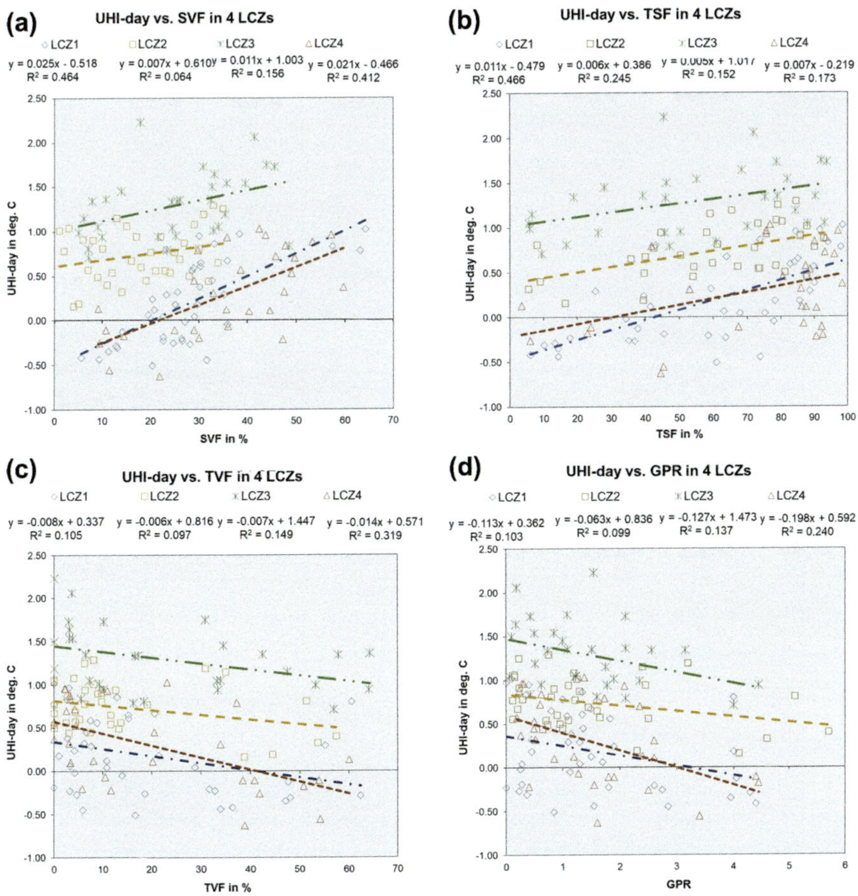

Fig. 4.15 Group-wise correlation of **a** TSF, **b** SVF, **c** TVF, and **d** GPR with UHI-day

GPR are negatively related to UHI-day in all UCZs. TSF meets the 5% significant level in all UCZs, and the R^2 values range from 0.15 to 0.47. The R^2 of SVF at LCZ 4 is very low, and correlation is not significant at the 5% level. The narrow range of SVF values in LCZ 4 (0–35%) could influence the result. Similarly, the low R^2 values of TVF and GPR at LCZ 4 could be attributed to the uneven distributions: The sites Nanxing and Shuxiang in LCZ 4 feature very low tree covers and leaf densities.

Scatter plots and correlations of GPR, GCR, and TVF with UHI-night are shown in Fig. 4.16. The correlations of GPR and GCR with UHI-night in LCZ 4 and the correlations of TVF with UHI-night in LCZs 3 and 4 are very weak, although a negative relationship is indicated in all the three plots. Similar to the UHI-day situation, narrow ranges and uneven distribution of the independent variables in the site of LCZ 4 can contribute to the low R^2 values.

Considering the correlation of UHI-night with SVF (Fig. 4.16d), data from both LCZs 2 and 3 indicates significant positive relationships. Negative relationships are

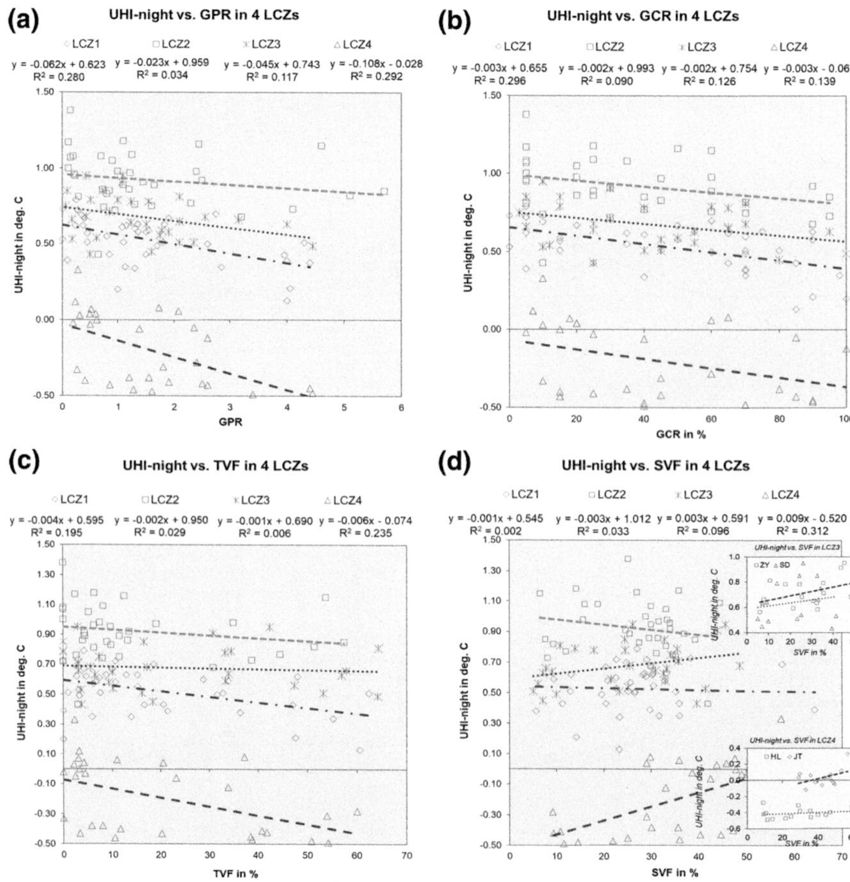

Fig. 4.16 Group-wise correlation of **a** GPR, **b** GCR, **c** TVF, and **d** SVF with UHI-night

detected in LCZs 1 and 4, but the null hypothesis about the zero correlation cannot be rejected because both models are failed to meet the 5% significant level. So in a statistical sense, the appeared negative relations can probably be caused solely by chance. Overall, the result indicates that a higher sky view tends to increase the UHI intensity at night. Theoretically, SVF is negatively related to maximum UHI (mostly occurred at night) as it controls the radiative cooling to the sky as a heat sink at night (Oke 1981). Oke cautioned, however, that this single measure may not be able to describe UHI patterns *in a particular city* even under "ideal" conditions, because of perturbation of other factors including thermal admittance, anthropogenic heat release, etc. (Arnfield 2003). The investigated residential quarters show varied ground surface materials and strong anthropogenic heat especially at night. From the perspective of energy balance, more open locations generally have higher solar uptake during the day; consequently, the heated surfaces tend to emit more heat at night and lift the air temperatures. Future experimental studies are needed to better understand the extent of impacts from above factors. But it seems clear that increasing the openness is not an effective way to improve the nighttime thermal condition in high-rise residential quarters in Shanghai.

Overall, none of the independent variables is capable of explaining a majority of variation in UHI intensity. The values of R^2 range from 0.11 to 0.46. In experimental studies on UHI, the bivariate correlation normally yields rather low R^2 values, e.g., a R^2 within 0.38–0.46 obtained in a study using SVF as the independent variable in Japan (Yamashita et al. 1986) and a R^2 of 0.28 in a study to correlate the aspect ratio (H/W) with nighttime UHI in Singapore (Goh and Chang 1999). This indicates that the UHI problem can be significantly related to but cannot be solved by a single design variable, as there are many factors contributing to the UHI formation in dense urban areas.

4.2.4 Multi-variant Regression at the Urban Scale

Multiple regression analysis incorporates all the independent variables. The Spearman correlation test is applied to examine significant relationship with dependent variables (at the 5% level, 2-tailed) and to detect any significant inter-correlation among independent variables. The significant but inter-correlated independent variables are entered into regression modeling separately. The three categorical variables are forced into all the models in order to check the background effect. The objective is to build more parsimonious models that exclude irrelevant independent variables and to reduce the impacts from inter-correlation among independent variables on the partial regression coefficients.

- UHI-day

The Spearman correlation indicates that Z1, Z2, Z3, SVF, TSF, TVF, GPR, GCR, WVD, SRS, and FAR (refer to Table 3.3) are significantly related to UHI-day at the 5% level. But correlations at various degrees occur among SVF, TVF, TSF, GCR, and GPR; SRS and FAR are also seriously correlated with each other. Following the aforementioned approach, the five inter-correlated point-specific variables (i.e., SVF, TVF, TSF, GPR, and GCR) and two site-specific variables (i.e., FAR and SRS) are entered into regression one pair at a time, in addition to the categorical variables Z1, Z2, and Z3. The total combination is $2 \times 5=10$ (Tables 4.6 and 4.7).

The explanatory powers (R^2) of the yielded models range 0.69–0.77. The models with FAR entered are higher in R^2 value than the models with SRS entered. The difference is in the order of 1–2%. This is because FAR is more closely related to UHI-day (Spearman correlation, $\rho = -0.349$) than SRS ($\rho = 0.232$). The coefficients indicate that, within the critical ranges, a 0.1–0.2 °C reduction in UHI-day is possible either by increasing the FAR by 1 or by reducing the SRS by 50 W/m^2. The partial coefficients of the point-specific variables suggest that, within the critical ranges (shown in Fig. 4.15), a 0.3 °C reduction in UHI-day is possible, by either lowering SVF by 20%, increasing TVF by 30%, lowering TSF by 50%, or increasing GPR by 2. In addition, increasing WVD by 1 m/s could reduce UHI-day by 0.3–0.4 °C.

The UHI-day differences induced by the background effect as indicated by the coefficients of Z1, Z2, and Z3 are in general agreement with the descriptive statistics. The background effect has a major impact on the daytime UHI, as the model including only three categorical variables (not shown here) explains more than half of the UHI-day variation ($R^2 = 0.528$). This means that adding other on-site design variables into regression models can only increase the R^2 in the order of 0.2–0.25.

- UHI-night

The Spearman correlation indicates that Z1, Z2, Z3, SVF, TSF, TVF, GPR, GCR, WVN, GR, and FAR are significantly related to UHI-night. It is found that correlations at various degrees occur among SVF, TVF, TSF, GCR, and GPR and between GR and FAR. Regression randomly chooses one variable from the five correlated point-specific variables (i.e., SVF, TVF, TSF, GPR, and GCR) and one from two correlated site-specific variables (i.e., GR and SRS) for entering the regression model. The total combinations are $2 \times 5=10$ (Tables 4.8 and 4.9).

The generated models have R^2 values ranging 0.86–0.9. The models with GR entered have higher R^2 value than the models with FAR entered. Comparing the two site-specific variables, it is found that although FAR is slightly stronger related to UHI-night ($\rho = 0.450$) than GR ($\rho = -0.447$), FAR is higher correlated to the categorical variables Z1, Z2, and Z3 ($\rho = 0.405$, -0.137, and -0.468, respectively) than GR does ($\rho = 0.112$, 0.028, and 0.229, respectively). The resulting overlapping of information conveyed by FAR causes the overall lower explanatory capacities of the models. The coefficients indicate that a 0.1–0.15 °C reduction in UHI-night is possible either by decreasing the FAR by 2 or by increasing the GR by 20%. The partial coefficients of the point-specific variables suggest that, within the critical ranges (shown in Fig. 4.16), a 0.1 °C reduction in UHI-night is possible, by either increasing GCR by 50%, increasing TVF by 25%, increasing GPR by 1.5, or increasing SVF by 20%. In addition, increasing WVN by 1 m/s could averagely reduce UHI-day by 0.1 °C. The coefficients of Z1, Z2, and Z3 are in general agreement with the descriptive statistics. The background effect has a dominant impact on the nighttime UHI.

Table 4.6 Regression results of the overall UHI-day models entering FAR and one of SVF, TVF, TSF, GPR, and GCR

Description/variables	Model-D1			Model-D2			Model-D3			Model-D4			Model-D5		
	Coefficients		Sig. level	Coefficients		Sig. level	Coefficients		Sig. level	Coefficients		Sig. level	Coefficients		Sig. level
	B	Beta		B	Beta		B	Beta		B	Beta		B	Beta	
Z1	−.612	−.475	.000	−.563	−.437	.000	−.607	−.472	.000	−.595	−.462	.000	−.524	−.407	.000
Z2	.393	.278	.000	.401	.284	.000	.403	.285	.000	.311	.220	.000	.339	.239	.000
Z3	−.724	−.506	.000	−.591	−.413	.000	−.653	−.456	.000	−.650	−.454	.000	−.672	−.469	.000
SVF	*.015*	*.336*	*.000*	–	–	–	–	–	–	–	–	–	–	–	–
TVF	–	–	–	*−.010*	*−.313*	*.000*	–	–	–	–	–	–	–	–	–
TSF	–	–	–	–	–	–	*.006*	*.296*	*.000*	–	–	–	–	–	–
GPR	–	–	–	–	–	–	–	–	–	*−.126*	*−.270*	*.000*	–	–	–
GCR	–	–	–	–	–	–	–	–	–	–	–	–	*−.004*	*−.188*	*.000*
WVD	−.309	−.249	.000	−.366	−.294	.000	−.321	.258	.000	−.351	−.282	.000	−.299	−.240	.000
FAR	−.132	−.281	.000	−.177	−.376	.000	−.124	−.263	.000	−.175	−.371	.000	−.198	−.421	.000
SRS	–	–	–	–	–	–	–	–	–	–	–	–	–	–	–
Constant	1.196		.000	1.986		.000	1.216		.000	2.013		.000	1.985		.000
R^2	.772			.769			.749			.748			.711		
Adjusted R^2	.762			.758			.737			.736			.697		
F static	71.80			70.37			63.23			62.79			52.04		
No. of cases	134			134			134			134			134		

Table 4.7 Regression results of the overall UHI-day models entering SRS and one of SVF, TVF, TSF, GPR, and GCR

Description/variables	Model-D6			Model-D7			Model-D8			Model-D9			Model-D10		
	Coefficients		Sig. level	Coefficients		Sig. level	Coefficients		Sig. level	Coefficients		Sig. level	Coefficients		Sig. level
	B	Beta		B	Beta		B	Beta		B	Beta		B	Beta	
Z1	−.658	−.511	.000	−.621	−.482	.000	−.651	−.505	.000	−.654	−.508	.000	−.589	−.457	.000
Z2	.389	.275	.000	.392	.277	.000	.406	.287	.000	.295	.209	.001	.328	.232	.000
Z3	−.760	−.531	.000	−.634	−.443	.000	−.666	−.466	.000	−.696	−.486	.000	−.714	−.499	.000
SVF	*.016*	*.360*	*.000*	–	–	–	–	–	–	–	–	–	–	–	–
TVF	–	–	–	*−.010*	*−.334*	*.000*	–	–	–	–	–	–	–	–	–
TSF	–	–	–	–	–	–	*.007*	*.319*	*.000*	–	–	–	–	–	–
GPR	–	–	–	–	–	–	–	–	–	*−.137*	*−.295*	*.000*	–	–	–
GCR	–	–	–	–	–	–	–	–	–	–	–	–	*−.004*	*−.197*	*.000*
WVD	−.322	−.259	.000	−.387	−.311	.000	−.334	−.268	.000	−.371	−.298	.000	−.316	−.254	.000
FAR	–	–	–	–	–	–	–	–	–	–	–	–	–	–	–
SRS	.003	.259	.000	.004	.360	.000	.002	.222	.000	.004	.355	.000	.004	.391	.000
Constant	.098		.519	.542		.001	.235		.140	.596		.000	.385		.028
R^2	.764			.755			.737			.735			.686		
Adjusted R^2	.752			.743			.724			.722			.672		
F static	68.35			65.22			59.20			58.57			46.34		
No. cases	134			134			134			134			134		

Table 4.8 Regression results of the overall UHI-night models entering FAR and one of SVF, TVF, TSF, GPR, and GCR

Description/variables	Model-N1			Model-N2			Model-N3			Model-N4			Model-N5		
	Coefficients		Sig. level	Coefficients		Sig. level	Coefficients		Sig. level	Coefficients		Sig. level	Coefficients		Sig. level
	B	Beta		B	Beta		B	Beta		B	Beta		B	Beta	
Z1	−.457	−.480	.000	−.417	−.437	.000	−.441	−.463	.000	−.459	−.481	.000	−.457	−.479	.000
Z2	−.270	.041	.000	−.254	−.243	.000	−.230	−.220	.000	−.226	−.216	.000	−.234	−.224	.000
Z3	−1.05	−.988	.000	−1.05	−.994	.000	−1.03	−.969	.000	−1.05	−.988	.000	−1.07	−1.01	.000
GPR	*−.067*	*−.196*	*.000*	–	–	–	–	–	–	–	–	–	–	–	–
GCR	–	–	–	*−.002*	*−.166*	*.000*	–	–	–	–	–	–	–	–	–
TVF	–	–	–	–	–	–	*−.004*	*−.164*	*.000*	–	–	–	–	–	–
TSF	–	–	–	–	–	–	–	–	–	*.002*	*.159*	*.000*	–	–	–
SVF	–	–	–	–	–	–	–	–	–	–	–	–	*.005*	*.138*	*.000*
WVN	−.118	−.127	.000	−.096	−.103	.004	−.113	−.122	.001	−.094	−.102	.005	−.087	−.094	.010
FAR	0.050	.144	.000	.037	.105	.007	.048	.138	.000	.070	.200	.000	.062	.177	.000
GR	–	–	–	–	–	–	–	–	–	–	–	–	–	–	–
Constant	.960		.000	.976		.000	.913		.000	.609		.000	.655		.000
R^2	.885			.875			.873			.869			.865		
Adjusted R^2	.879			.869			.867			.863			.859		
F static	162.60			147.71			145.66			140.86			135.62		
No. cases	134			134			134			134			134		

Table 4.9 Regression results of the overall UHI-night models entering GR and one of SVF, TVF, TSF, GPR, and GCR

Description/variables	Model-N6			Model-N7			Model-N8			Model-N9			Model-N10		
	Coefficients		Sig. level	Coefficients		Sig. level	Coefficients		Sig. level	Coefficients		Sig. level	Coefficients		Sig. level
	B	Beta		B	Beta		B	Beta		B	Beta		B	Beta	
Z1	−.373	−.391	.000	−.350	−.367	.000	−.360	−.377	.000	−.349	−.366	.000	−.359	−.376	.000
Z2	−.248	−237	.000	−.233	−.222	.000	−.211	−.201	.000	−.204	−.195	.000	−.214	−.205	.000
Z3	−1.04	−.982	.000	−1.04	−.982	.000	−.102	−.964	.000	−1.05	−.988	.000	−1.07	−1.01	.000
GPR	*−.065*	*−.187*	*.000*	–	–	–	–	–	–	–	–	–	–	–	–
GCR	–	–	–	*−.002*	*−.155*	*.000*	–	–	–	–	–	–	–	–	–
TVF	–	–	–	–	–	–	*−.004*	*−.154*	*.000*	–	–	–	–	–	–
TSF	–	–	–	–	–	–	–	–	–	*.002*	*.155*	*.000*	–	–	–
SVF	–	–	–	–	–	–	–	–	–	–	–	–	*.004*	*.127*	*.000*
WVN	−.109	−.117	.000	−.088	−.095	.005	−.103	−.112	.002	−.085	−.092	.007	−.077	−.084	.016
FAR	–	–	–	–	–	–	–	–	–	–	–	–	–	–	–
GR	−.007	−.170	.000	−.006	−.145	.000	−.006	−.167	.000	−.008	−.218	.000	−.008	−.195	.000
Constant	1.352		.000	1.283		.000	1.294		.000	1.143		.000	1.135		.000
R^2	.895			.885			.884			.883			.877		
Adjusted R^2	.890			.879			.878			.877			.0871		
F static	181.06			162.33			160.93			159.77			150.44		
No. cases	134			134			134			134			134		

The three categorical variables can explain more than 80% of UHI-night variation ($R^2 = 0.832$), while adding other on-site design variables into the models can only increase R^2 by 0.03–0.07. However, the R^2 values of models excluding the three categorical variables and including only the on-site design variables range 0.22–0.27. Clearly, there is overlapping between the on-site design variables and categorical variables due to collinearity.

4.3 Discussion

Climate consideration is the essential part of an environmental urban design if the goal is to provide comfortable and energy-conservative urban environment (Golany 1996). Urban design is about making choices, that is, to balance among a variety of social–cultural, economic, and environmental factors. Even considering the impact of urban climatology alone for a given climate zone, there are variations in urban form that favor a certain aspect of practical requirement (Oke 1988). In this light, the discussion on the study results to mitigate summertime outdoor UHI should be related to and balanced with other important environmental aspects in urban design, such as solar/daylight control, indoor environment, and streetscape.

Overall, the highest measured site-mean UHI reached 1.4 °C at day and 1.1 °C at night. Note that the strongest UHI in Shanghai occurs in late autumn and winter months, due to calm and clear weather conditions (Yan and Xu 1996a). However, although relatively weaker, the summertime UHI is most detrimental in terms of human comfort and energy efficiency. In this connection, the result of this study suggests that it be appropriate to set the threshold more stringently, e.g., 1 °C instead of 1.5 °C, for evaluation of summertime UHI in the urban context of Shanghai.

4.3.1 Site Planning Factors: Layout and Density

The impacts of building layout and density on UHI are different between day and night. This is largely determined by the impacts of the presence or absence of solar radiation. The interspersed tall buildings in Site 2 interweave with each other into a super umbrella over the open spaces within it. The trees on site with tall and dense canopies serve as the second barrier against the midday fierce sunshine. As a result, a daytime cool island is formed (Fig. 4.2). In contrast, the long-linear buildings in Site 3 shape the E–W-oriented canyons. Without enough tree shading, this layout suffers the highest UHI. At night, the reduced sky view by buildings and trees in Site 2 should be the main contributor to its highest UHI while anthropogenic heat should be influential too, due to its high density. Overall, Site 1 exhibits a good performance both at day and at night: It is slightly higher than Site 2 in UHI-day but becomes the coolest at night (Fig. 4.2). The semi-enclosed layout (supplemented by tree canopy) provides fairly good on-site shading.

Intensive land use is considered in agreement with sustainable principles (Wang et al. 2006) and also in line with the market interest. Practically in Shanghai, urban housing density is closely interrelated to spacing control and layout, while housing spacing is influenced by a series of factors including indoor daylight/solar admittance, fireproof and site ventilation, etc. (Li 2005).

Four types of building layouts are covered in this study (Fig. 1.4). In general, the interspersed layout geometry has the least mutual shading impact; consequently, the interspacing can be minimized and the highest density achieved, as shown in Site 2. In contrast, the long-linear layout as shown in Site 3 is the lowest in land-use efficiency. From the thermal perspective, high density in interspersed type is beneficial at day due to a higher shading factor but night situation can be worsen due to reduced SVF and intensified man-made heat. On the other hand, about a half of the units in this layout type do not have any south-facing rooms, which have the most comfortable indoors during both summer and winter in Shanghai's climate (Yan and Xu 1996b). The linear type is the most widely accepted one for its capability of harnessing the natural ventilation and solar energy for passive cooling/heating. In winter, the N–S orientation enables low-angle solar access to southern indoor spaces, while the tall buildings on the northern site boundary block the prevailing northern winds. In summer, the gradually lowered buildings toward the southern boundary allow the site to open toward the prevailing southeastern winds, thereby enhancing natural ventilation (Givoni 1998). Further, for a given density, a careful mix of buildings of various heights helps improve local ventilation as compared to a homogeneous height setting (Ng et al. 2006). One of the drawbacks of the linear form can be its relatively lower density. The study also indicates that, although this layout has an overall good nighttime thermal condition, it tends to suffer high daytime UHI due to high solar exposure (e.g., Yishan in LCZ 1 and Shuxiang in LCZ 2). This issue can be addressed with carefully planted trees within the east–west-oriented canyons or by adding east–west-facing buildings at the ends of the canyons to provide further shadow benefit. This resultant semi-enclosed layout also increases the density which is economically sound, but the addition should be carefully managed to avoid negative impact on the site ventilation. Therefore, considered in a holistic way, the semi-enclosed layout is the best choice because of the good UHI mitigation performance, southern orientation for most of the units, and a fairly high density. In addition, it contributes to a more integrated urban streetscape compared to interspersed and linear layouts. Among them, bioclimatically, the semi-enclosed layout is analogous to Yeang's green skyscraper strategy which locates service core at west–east facades to avoid overheating of salable space (Yeang 1994). The east–west-facing buildings in this scenario should have specific shading details and proper built form to control summertime insolation.

4.3.2 Greenery Parameters: GR, TVF, GPR, and GCR

A green ratio of 35% is mandatory for all new residential developments in Shanghai (Shanghai Municipality 2000), which is 5% higher than the national standard (MOHURD (Ministry of Housing and Urban-Rural Development) 2016). This study, however, indicates that increasing GR does not necessarily lead to lowered daytime UHI, because site shading factor could dilute or negate vegetative cooling performance. For instance, the extensive shadow cast by tall buildings (e.g., in Xiling and Nanxing) nullified the cooling effect of green cover to a great extent. In contrast, high daytime UHIs were recorded on unshaded open lawns (e.g., at Yishan and at Shuxiang). GR appears to be negatively correlated with UHI-night, but it is difficult to determine whether the cooling effect is caused by lower density or higher green cover, because of the collinearity of FAR and GR. This kind of problem is difficult to avoid in an empirical study. Nevertheless, for all practical purposes, the two variables can be manipulated collaboratively to mitigate the nighttime UHI effect (i.e., to reduce building volume and consequently make more room for greenery). On the other hand, leaf density cannot be measured by GR which is a two-dimensional indicator. As such, the green plot ratio (GPR) should be extended to the site level in order to better quantify greenery biomass (Ong 2003).

For the three point-specific greenery variables: TVF, GPR, and GCR, regression analysis reveals that TVF and GPR are significantly related to daytime UHI, while GPR and GCR significantly related to nighttime UHI in most data LCZs. In addition, TVF appears negatively related to nighttime UHI intensity in all data LCZs. The findings suggest several implications in landscape design:

1. *Increase site leaf density at major open spaces in housing quarters by planting trees with tall trunks and dense canopies.* It is found that green space at most high-rise housing quarters in Shanghai features extensive grass cover. Shrubs and trees are dotted only for decorative purposes. The study shows that substantial cooling effects could be realized by increasing GPR to the range of 3–4. The effort to increase green biomass should be focused on increasing tree cover, as the study shows that increasing TVF to 40–50% produces tangible cooling effects. A minimum of three arbor trees is required per 100 m^2 of residential green space, to get one credit in the evaluation standard (MOHURD (Ministry of Housing and Urban-Rural Development) 2006). It would be better to clearly stipulate the minimal plant height and canopy size, and provide more credits for extra trees.
2. *Design amenity spaces that are well shaded by buildings.* Psychologically, dense green spaces may trigger a perception of fear to potential crime or hidden animals and thus prevent active use of the space (Yokohari et al. 2006). Therefore, an "open" space covered with a lawn or hard pavement remains an important amenity for outdoor activities. The location of open spaces should be carefully considered with respect to solar geometry and on-site building layout: It should be placed under the shade cast by buildings during the major periods of use during summer

days (e.g., from sunrise to 10 am and from 4 pm till sunset). Site shading factor can be evaluated at the design stage with the aid of simulation tools.

3. *Leave sufficient and carefully organized gaps for ventilating outdoor spaces where shrubs are applied.* When wind speed exceeds 1 m/s, ventilative cooling is an effective way to lower both daytime and nighttime UHI intensities. However, dense vegetation, particularly low shrubs and bushes, can act as wind barriers and thus aggravate thermal discomfort. Generally, an outdoor space should open to the south to introduce southerly breezes in summer, whereas dense shrubs and bushes can be planted at the northwestern boundary to block cold winter winds.

4.3.3 Radiation Exchange: TSF, SVF, and GSA

Solar radiation is the determining factor in daytime UHI intensity. TSF used in this study well quantified the solar access, considering the urban geometry and solar geometry in an integrative manner. Notably, TSF is calculated based on clear sky conditions (i.e., the diffuse/direct radiation ratio is set at 0.15), but it has a higher explanatory power in daytime UHI than SVF even under a partially cloudy sky condition. For data from LCZ 4 that was measured under a clear sky, TSF is the only statistically significant variable. Overall, TSF shows a consistently positive relationship to daytime UHI in all LCZs, significantly at the 5% level. TSF thus has the potential to be used as an indicator in outdoor thermal assessment and urban design guidelines.

During the day under partially cloudy conditions, SVF is a good indicator of solar access and thus daytime UHI (Fig. 4.17). At night when the radiation exchange between the sky and ground reverses, theoretically SVF should be negatively related to nighttime UHI. However, the LCZ-wise analysis indicates an overall positive relationship. This is in agreement with Oke's caution on his UHI formula that it would underestimate UHI in situations where anthropogenic heat releases are large (Oke 1987). Therefore, building layout and landscape design should be collaboratively manipulated to maintain a reasonably low SVF level, without sacrificing site ventilation potential.

The study reveals that increasing ground surface albedo alone does not have any significant impact on UHI intensity without considering shading factors. On the local microscale, increasing albedo values of the ground or building façade in a high-rise built environment can exacerbate daytime thermal discomfort due to the high amount of reflected radiation, and this reflected solar heat may cause indoor glare and increase cooling energy use. As an alternative, permeable pavement could be applied in high-rise residential developments to effectively lower UHI without causing the previously mentioned problems.

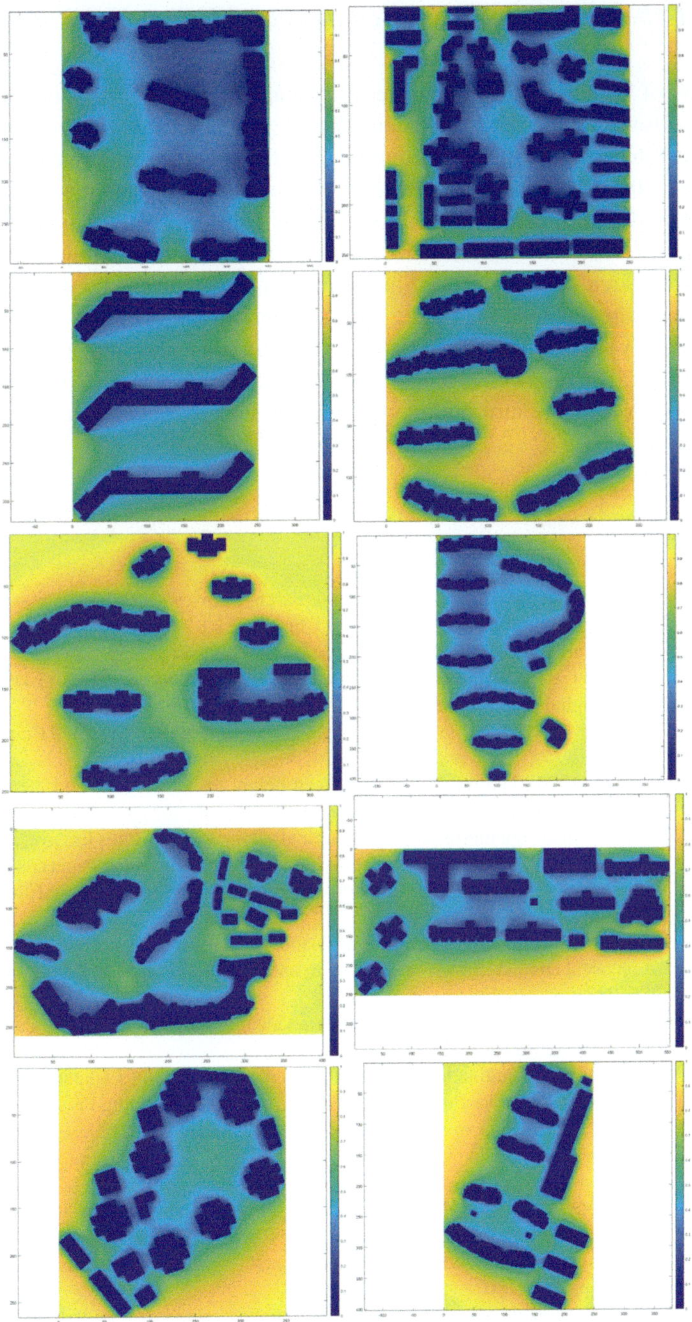

Fig. 4.17 SVF distributions at the ten experimental sites computed based on DEM

4.3.4 Urban-Scale Factors: Regional and Urban Ventilation

The focus of the case study is on the microscale thermal effort of on-site design variables. However, the results indicate that the problem cannot be solved without considering the urban-scale factors. Future research is needed to derive urban design strategies on the urban scale. Nevertheless, the findings suggest that ventilation conditions should be improved on the city scale by considering the prevailing winds. The range for which the sea influences temperature is approximately within 20 km of the coastal area (Givoni 1998). Shanghai inner city is on the border of that line. Possible measures include regulating building density and planning wind corridors to facilitate city ventilation. The measures are useful even when the prevailing winds are very weak, because on calm and clear nights, the buoyancy-driven centripetal flow of cool air from the surrounding rural areas to the urban center can be enhanced.

References

Arnfield AJ (2003) Two decades of urban climate research: a review of turbulence, exchanges of energy and water, and the urban heat island. Int J Climatol 23:1–26

Chow SD (1982) The urban heat island of Shanghai. Acta Geogr Sin 37:372–381

Ding J, Zhang Z, Xi H (2002) A study of the high temperature distribution and heat island effect in the summer of Shanghai area. Chinese J Atmos Sci 26:412–420

Givoni B (1998) Climate considerations in building and urban design. Van Nostrand Reinhold, New York

Goh KC, Chang CH (1999) The relationship between height to width ratios and the heat island intensity at 22:00 h for Singapore. Int J Climatol 19:1011–1023

Golany GS (1996) Urban design morphology and thermal performance. Atmos Environ 30:455–465

Hoyano A, Iino A, Ono M (1999) Analysis of the influence of urban form and materials on sensible heat flux—a case study of Japan's largest housing development "Tama New Town". Atmos Environ 33:3931–3939

Kolokotroni M, Giridharan R (2008) Urban heat island intensity in London: an investigation of the impact of physical characteristics on changes in outdoor air temperature during summer. Sol Energy 82:986–998

Li Z-Y (2005) A discussion on the sunlgiht rule for residential building in the city center. Urban Plann Forum 155:79–82 (in Chinese)

MOHURD (Ministry of Housing and Urban-Rural Development) (2006) Evaluation standard for green building. GB 50378-2006. Beijing, China Architecture and Building Press

MOHURD (Ministry of Housing and Urban-Rural Development) (2016) Design and planning code for urban residential sectors. GB 50180-93 2016 *Edition*. Beijing, China Architecture and Building Press

Ng E, Chan T-Y, Cheng V, Wong NH, Han M (2006) Designing high-density cities—parametric studies of urban morphologies and their implied environmental performance. In: Bay J-H Ong BL (eds) Tropical sustainable architecture: social and environmental dimemsions. Architectural Press Oxford, UK, pp 151–180

Oke TR (1981) Canyon geometry and the nocturnal urban heat island: comparison of scale model and field observations. Int J Climatol 1:237–254

Oke TR (1987) The boundary layer climates, 2nd edn. Methuen, London and New York

Oke TR (1988) Street design and urban canopy layer climate. Energy Build 11:103–113

Oke TR (2004) Initial guidance to obtain representative meteorological observations at Urban Sites: IOM Report No.81, WMO/TD. No. 1250. World Meteorological Organization Geneva, Switzerland

Ong BL (2003) Green plot ratio: an ecological measure for architecture and urban planning. Landscape Urban Plann 63:197–211

Shanghai Municipality (2000) Regulation on urban landscaping in urban shanghai. Shanghai Municipality

Wang Y, Han J, Zeng J, Yang J, Fang T (2006) Evaluation standard for green building. China Architecture and Building Press, Beijing

Yamashita S, Sekine K, Shoda M, Yamashita K, Hara Y (1986) On relationships between heat island and sky view factor in the cities of Tama River basin, Japan. Atmos Environ 20:681–686

Yan J, Xu J (1996a) Shanghai climate. Meteorology Press, Beijing, China, (in Chinese)

Yan J, Xu J (1996b) Shanghai climate. Meteorology Press, Beijing, China

Yang F, Lau SSY, Qian F (2010) Summertime heat island intensities in three high-rise housing quarters in inner-city Shanghai China: building layout, density and greenery. Building Environ 45(1):115–134. https://doi.org/10.1016/j.buildenv.2009.05.010

Yang F, Lau SSY, Qian F (2011) Urban design to lower summertime outdoor temperatures: an empirical study on high-rise housing in Shanghai. Building Environ 46(3):769–785. https://doi.org/10.1016/j.buildenv.2010.10.010

Yeang K (1994) Bioclimatic skyscrapers (with essays by Alan Balfour and Ivor Richards). Artemis, London

Yokohari M, Amemiya M, Amati M (2006) The history and future directions of greenways in Japanese New Towns. Landscape Urban Plann 76:210–222

Chapter 5
Pedestrian Wind in High-Rise Residential Quarters

Abstract *Study intent* Urban wind environment has significant impacts on pedestrian comfort and safety, as well as pollution dispersion and building energy consumption. This case study empirically examines the microscale effect of urban form, density and greenery on summertime and wintertime outdoor pedestrian-level wind environment and users' thermal and wind comfort in selected high-rise residential neighborhoods in Shanghai. Based on the findings, proper design strategies are formulated for high-rise residential neighborhoods in the so-called hot-summer cold-winter (HSCW) climate zone in China, so as to optimize ventilation potential in hot-humid summer and transient seasons while protecting users from strong winds in cold season. *Results and discussion* In summer, the wind statistics indicates significant influence from urban geometry. The pedestrian-level wind velocity ratio (WVR) is significantly correlated with the "degree of enclosure" indicated by sky view factor (SVF), due to buildings and/or greenery. It suggests that within the practical range, increasing SVF by 10% could increase WVR by 7–8%. Under the observed weak wind environment, SVF could indicate the thermal buoyancy-driven airflow rate that is determined by solar radiation heating. At the site level, WVR is significantly negatively correlated with building volumetric density (floor area ratio, FAR). In the winter survey, the measured WVR is highest in the long-linear high-rise building layout, and is lowest in the mid-rise linear building layout; but the questionnaire survey reveals that there are still 87% of respondents felt fairly comfortable in the long-linear high-rise building layout, only 7% less than the mid-rise building layout. Overall, since winter wind is not harsh, summer ventilation should be prioritized in high-rise building design in Shanghai considering seasonal prevailing wind directions. Winter shelter, if needed, can be provided by proper landscape design.

This chapter is adapted from published journal papers, Yang et al. (2013) and Yang (2015).

F. Yang and L. Chen, *High-Rise Urban Form and Microclimate*,
The Urban Book Series, https://doi.org/10.1007/978-981-15-1714-3_5

5.1 The Summer Period

5.1.1 Introduction

This study aims to measure the microscale wind effect of site planning and building design in summertime through an empirical approach. The microscale wind availability, and thus the extent of wind impact of urban form, is represented by the wind velocity ratio (WVR), the ratio of the wind measured at the pedestrian level to that at a pre-determined reference point. The position of the reference wind sensor should be carefully located, with which the pedestrian-level wind velocity can be compared. According to the theory of scales in urban climatology, this reference point should be able to "see" the local-scale urban effect and is deemed free of the microscale influence. It can be put vertically above the urban area of interest, over a critical height termed blending height (Z_r) that separates roughness sub-layer (RSL) and inertial sub-layer (ISL). Depending on the local urban geometrical properties, Z_r can be ranged 1.5–4 times average canopy height (Z_H) (Oke et al. 2017). However, for high-rise building clusters in a heterogeneous surrounding urban environment, ISL may not exist at all. Even if it exists, to put a sensor stably over four to five hundred meters above in the sky does not seem feasible. An alternative location is far upstream, at the mean building height of the working site (Bottema 1999). Knowing the background wind at an ideal meteorological station, we can use this reference to separate the city-scale wind influence from the building-scale influence (Blocken and Carmeliet 2004).

The reference wind is approximated here based on the idea of local climate zone (LCZ) that also applied in the UHI study (Chap. 4). Among the factors in LCZ clustering, roughness class, aspect ratio, and built ratio are directly related to ventilation potential. We have classified the inner city of Shanghai into four LCZs according to urban cover and urban structure (Chap. 4). The variation of local wind availability within each LCZ is assumed marginal. The details of the classification process can be found in the previous chapter. Topographically, the terrain-related variation is assumed marginal among all the measurement sites because Shanghai is a flat city with negligible altitude variation. The dependent variable WVR in this study is defined as the on-site mean wind speed measured at the pedestrian level, to the mean wind speed at the windward-side rural observatory. Chuansha Observatory was chosen as the reference station, because it is located to the windward side of the inner-city area, under the prevailing Southeastern summer winds (Fig. 5.1).

5.1.2 Results

The applied datasets are from the same field work that produced the data for the UHI study (Chap. 4). The wind data includes 36 Hobo weather stations from the ten sites that are grouped into four LCZs (Fig. 4.2). The wind data is measured from 1 to

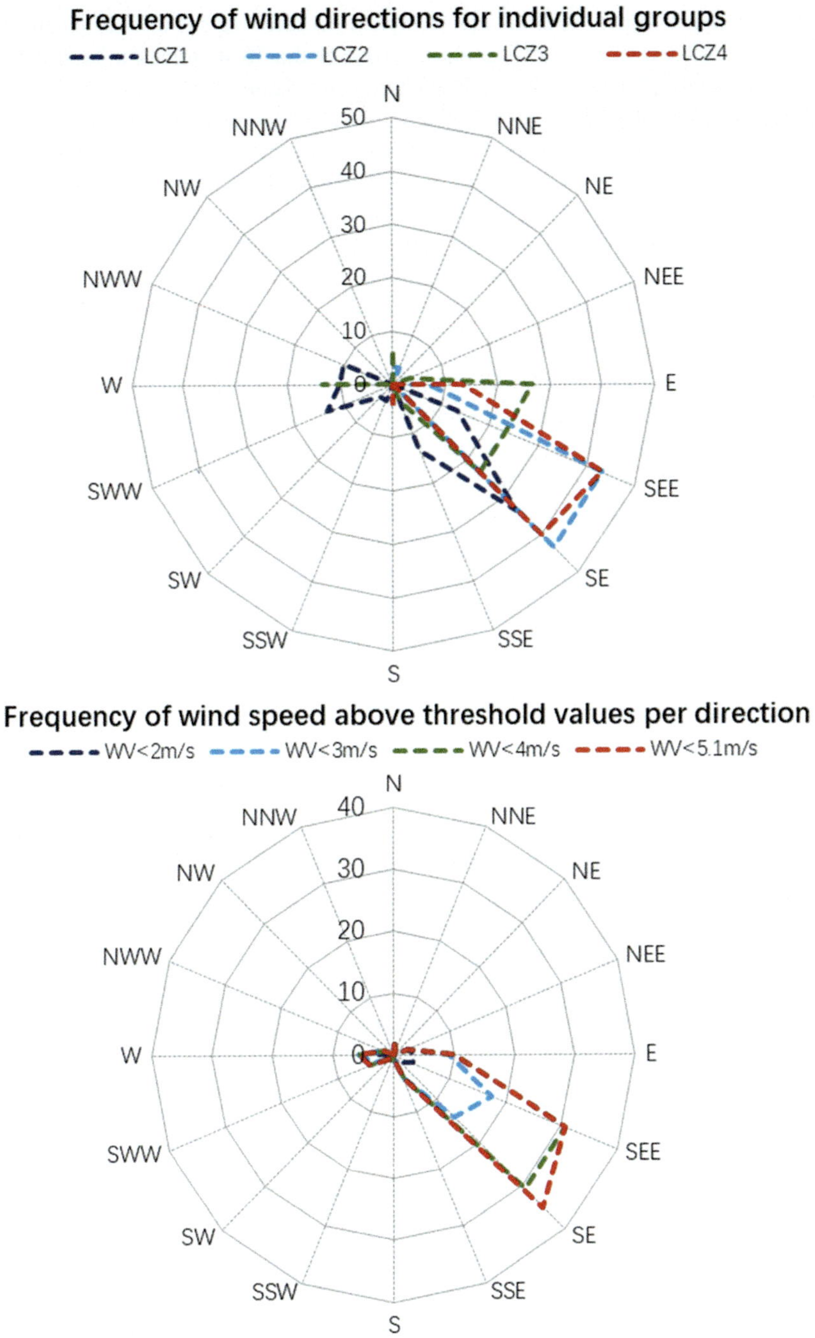

Fig. 5.1 Frequency of wind direction and speed per direction during the measurement period at the Chuansha reference station

9:30 pm at a one-minute interval, consecutively for three days (Fig. 5.2). Therefore, the number of datasets at each station is 510. For the 36 stations, the total dataset number is 18,360. Three days' data is averaged and clustered into daytime (1–6 pm) and nighttime (6–9:30 pm) period. The analysis firstly examines the distribution of wind velocity (WV) against air temperature (Ta) on the LCZ level, and plots it against a predefined outdoor comfort zone. Then, site-mean WV and WVR values are computed and correlated against the form and density variables, and site wind direction (WD) plots are qualitatively analyzed. Thirdly, WVR values are correlated with the microscale form and density variables (of building and/or greenery).

- WV and Ta at the zone level

The environmental factors influencing human comfort include air temperature, wind velocity, radiant temperature, and humidity. Research conducted in similarly hot-humid regions (Japan and part of Israel) has indicated that RH is statistically insignificant in influencing outdoor thermal comfort perception (Givoni et al. 2003). According to the empirical formula developed by Givoni et al., within certain biometeorological range, an increase of 0.35 m/s in wind speed has equivalent effect to a 1 °C decrease in air temperature (Givoni et al. 2003). Solar radiation is constant changing in outdoor environment, and the measurement of the radiant temperature requires sophisticated equipment. Since it is common sense for Southeast Asian people that in summer days one can hardly be thermally comfortable under direct sunshine, so to map the comfort zone in a simple way, this study assumes a shaded situation for all stations during the daytime measurement period. The box plots of WV by Ta of each LCZs are given in Fig. 5.3. A comfort zone is overlapped on the plots, based on the summarized outcomes from studies on outdoor thermal comfort in Japan, Israel and Hong Kong (Givoni et al. 2003; Ng 2008). The zone is only a general outline of the comfort range that applies under certain conditions (i.e., 27 °C < Ta < 34 °C; WV < 5 m/s; clothing level: ~0.65; activity level: sitting, standing or walking). It is intended to show the approximate comfort zone in relation to plotted data. It is not an accurate delineation of the comfort boundary, in that it does not consider the two-degree variation within each Ta category. The exact number of datasets that meet the comfort criteria is tabulated under each plot.

It is clearly seen from the plotted data that, thermal comfort is hardly achieved during most of the time, at most of the measured locations. The LCZ1 and LCZ2 were both measured during the peak summer day period. In LCZ1, no datasets meet the comfort criteria, and only ten datasets in LCZ2 enter the comfort zone. In LCZ1, 56% of the datasets have Ta values that are higher than 34 °C, meaning that due to high air temperatures, comfort is no longer achievable by solely increasing WV. During the late summer day period, there are 0.5% and 6% of the measured data that enter the comfort zone, from LCZ3 and LCZ4, respectively. The qualified datasets were all measured during the nighttime (6 pm–9:30 pm). Note that the WV mean, median, and variability (standard deviation) all tend to increase with the increase in Ta, suggesting that thermal buoyancy is the major force behind the measured air movement. However, the effect is obviously not enough to reach the comfort zone.

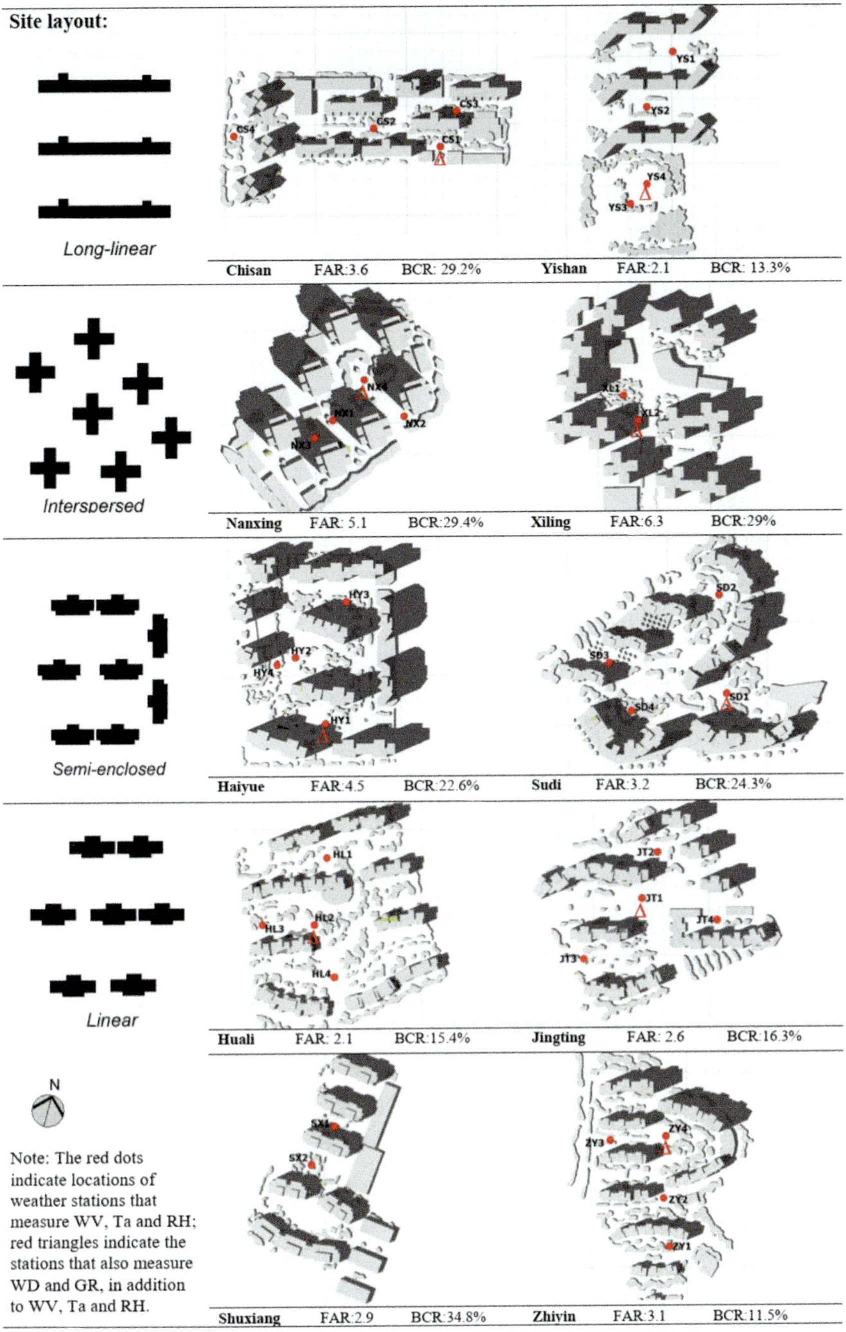

Fig. 5.2 Layout and properties of ten sites and the wind measurement points

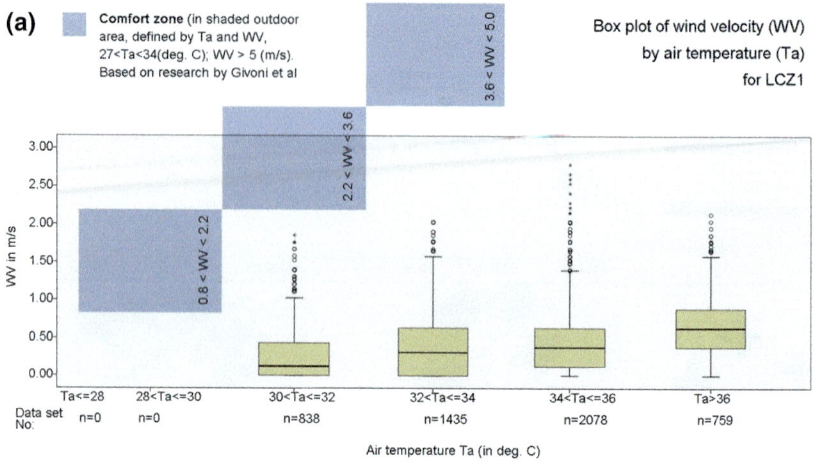

WV stats/Ta range (in °C)	28-30	30-32	32-34	34-36	>36
Frequency	-	838	1435	2078	759
mean	-	0.29	0.41	0.42	0.66
median	-	0.13	0.31	0.38	0.63
SD	-	0.37	0.41	0.42	0.42
Upper quartile	-	0.43	0.63	0.63	0.89
Lower quartile	-	0	0	0.13	0.38
In comfort zone	-	0	0	0	0

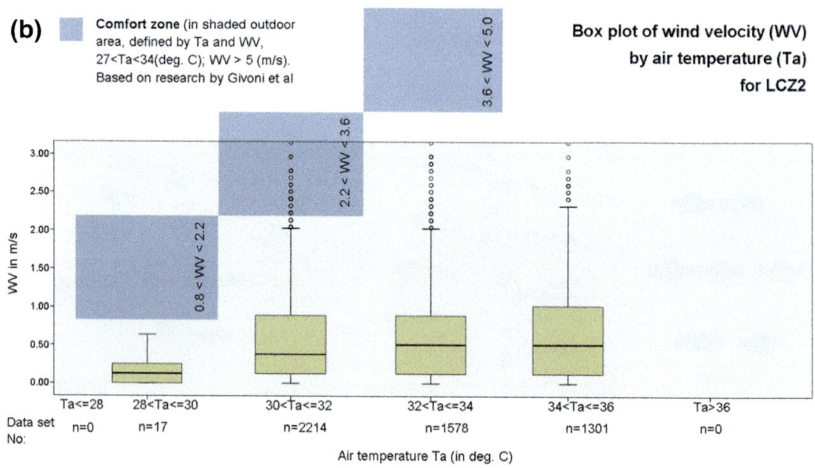

WV stats/Ta range (in °C)	28-30	30-32	32-34	34-36	>36
Frequency	17	2214	1578	1301	-
mean	0.13	0.55	0.59	0.64	-
median	0.13	0.38	0.51	0.51	-
SD	0.18	0.55	0.56	0.60	-
Upper quartile	0.25	0.89	0.89	1.01	-
Lower quartile	0	0.13	0.13	0.13	-
In comfort zone	0	10 (0.2%)	0	0	-

Fig. 5.3 Box plots of wind velocity by air temperature for sites from four LCZs

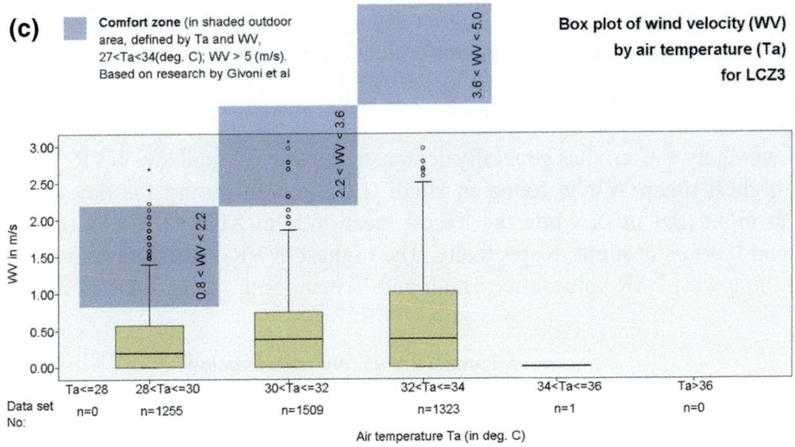

WV stats/Ta range (in °C)	28-30	30-32	32-34	34-36	>36
Frequency	1255	1509	1323	1	-
mean	0.34	0.46	0.63	0	-
median	0.19	0.38	0.38	-	-
SD	0.45	0.52	0.65	-	-
Upper quartile	0.57	0.75	1.02	-	-
Lower quartile	0	0	0	-	-
In comfort zone	23 (0.5%)	0	0	0	-

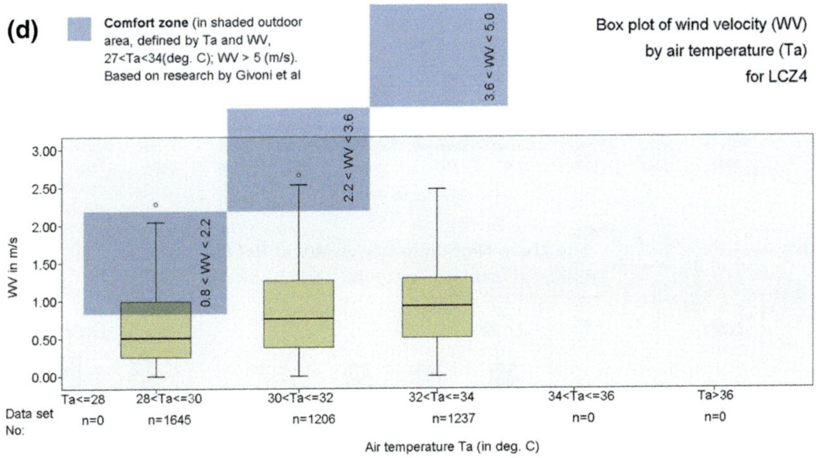

WV stats/Ta range (in °C)	28-30	30-32	32-34	34-36	>36
Frequency	1645	1206	1237	-	-
mean	0.62	0.86	0.96	-	-
median	0.51	0.76	0.93	-	-
SD	0.48	0.54	0.50	-	-
Upper quartile	0.99	1.27	1.30	-	-
Lower quartile	0.25	0.38	0.51	-	-
In comfort zone	247(6%)	0	0	-	-

Fig. 5.3 (continued)

Overall, the measured data reveals a hot and very weak outdoor wind environment, which contributes to thermally uncomfortable outdoor spaces.

- Wind effect of building form and density at the site level

Site-averagely, the ten sites generally demonstrated low WV and low WVR (Fig. 5.4). The highest mean WV is found in Huali (LCZ4) both during the day (1.2 m/s) and at night (0.9 m/s), while the lowest mean WV in Xiling (LCZ1), 0.4 m/s at day and 0.3 m/s at night, respectively. The highest WVR values are found at sites with the lowest FAR values, i.e., Huali and Yishan, both with a FAR of 2.1, while

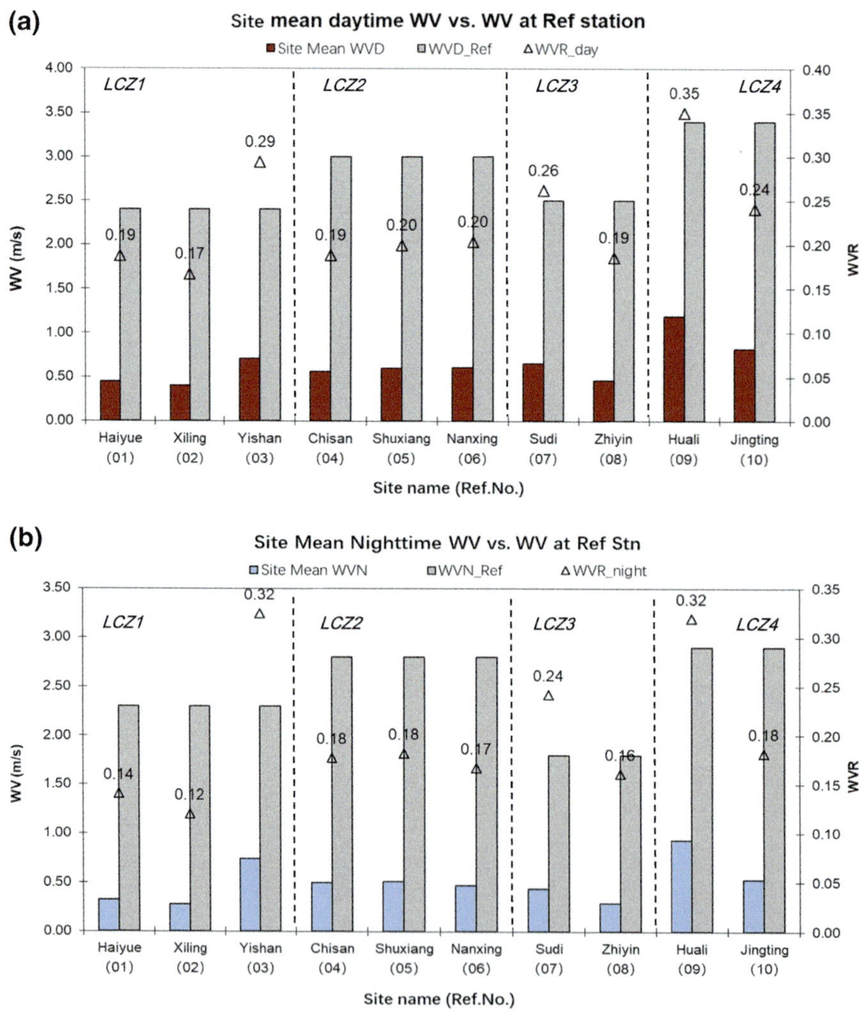

Fig. 5.4 Site-mean **a** day and **b** night wind velocities, reference wind velocities and WVRs of ten sites

sites suffering from the lowest WVRs feature a very high FAR range of 4.5–6.3, i.e., Haiyue and Xiling (Fig. 5.2). The comparison indicates a negative relationship between WVR and FAR. No clear relation is found between WVR and site layout. Zone-averagely, sites in the LCZ4 demonstrated the best ventilation performance. Differences among the rest three zones are not very significant, with the sites in the LCZ2 exhibits a little lower WVR compared with LCZ1 and LCZ3. There can be two reasons: First, the FAR value of LCZ4 is the lowest at the zone level as well as at the site level; second, under the prevailing southeast winds (Fig. 5.1), the LCZ4 is on the windward side to other zones, and consequently enjoys the least wind velocity loss due to surface drag.

A scatter plot and correlation test show a strong and significant correlation between WVR and FAR, both during the day ($\gamma = -0.697$, Sig. level < 0.05, 2-tailed) and at night ($\gamma = -0.729$, Sig. level < 0.05, 2-tailed) (Fig. 5.5a). Although a negative relationship appears between BCR and WVR (Fig. 5.5b), the relationships are insignificant neither during the day nor at night, at the 5% level. Note that this bivariate analysis does not take the vegetation effect into account, therefore the determination of exact correlation coefficiencies of the models still require further work.

The WD frequency statistics at selected open spaces at eight of the ten sites are shown in Fig. 5.6. All of the WD patterns have strong directional characteristics, but none of them conforms to the WV pattern at the reference rural station (Fig. 5.1), indicating strong impact from the surrounding built environment.

For the long-linear sites Yishan and Chisan, the recorded winds are predominantly from north and northwest (Fig. 5.6b and c), which are opposite to the southeast direction at the reference station. In both sites, the WD measurement points were located to the windward side of long slab tall buildings (CS1 in Chisan and YS4 in Yishan, Fig. 5.2). The building façade can divert a part of the incoming airflow downward and produce a vortex near the ground surface (Oke 1987). The measured WD pattern could indicate the reverted airflow near the ground. For the interspersed sites, i.e., Nanxing and Xiling (Fig. 5.2), the dominant WD patterns correspond to the gaps between surrounding individual high-rise towers (Fig. 5.6a and d). For the two linear sites, Huali and Jingting, in the same LCZ (the ER zone), the measured WD patterns are similarly northeastern, largely reflecting the opening directions at measurement points (Fig. 5.6e and f). The WD pattern at Sudi perfectly fits the northeast opening of the surrounding building envelope, demonstrating a strong channeling effect by the tall buildings. Sudi and Zhiyin are both located to the east of the Suzhou Creek. The eastern winds measured at Zhiyin probably indicate the influence of the thermal-induced air movement from the creek to the adjacent land area. The effect was not observed at Sudi probably because of the walling effect of the long buildings and the elevated pedestrian walkway in front of the creek (Fig. 5.6g and h).

- Wind effect of form and density at the micro scale

Statistical relationships are examined between ventilation potential and the site enclosure factor. The WVR as defined in previous sections is used as the dependent variable to quantify the site ventilation potential. Three independent variables are used

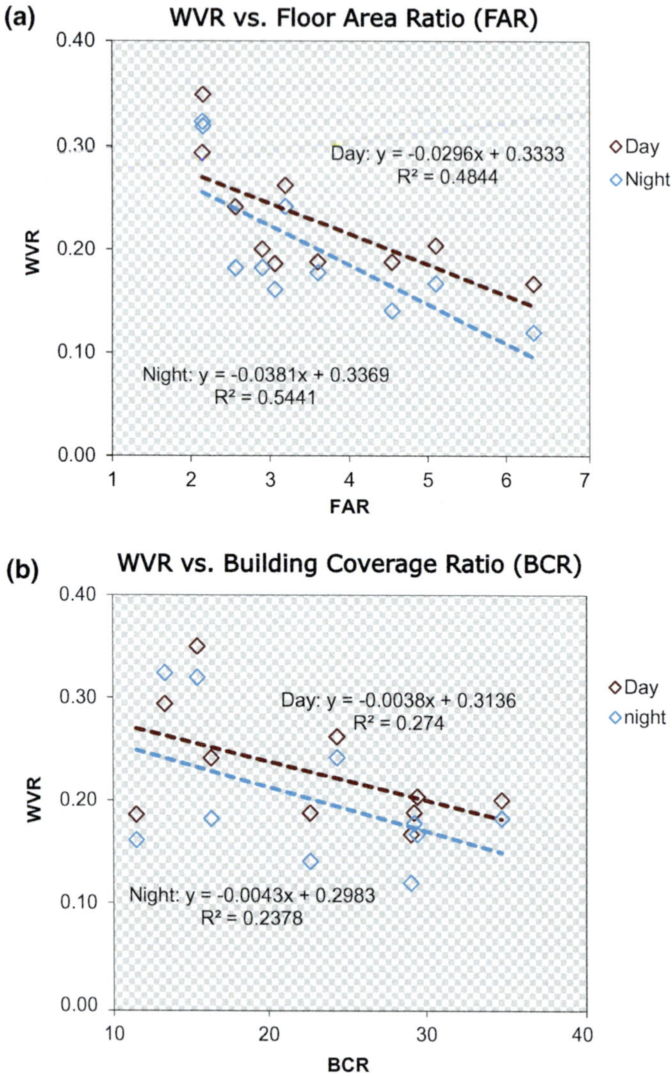

Fig. 5.5 Scatter plots and linear-fit of **a** FAR with WVR, and **b** BCR with WVR

to quantify the enclosureness: sky view factor (SVF), taking into account of the integrated effect of building and vegetation canopies; tree view factor (TVF), considering only the shelter impact from the overhead vegetation; and green plot ratio (GPR), considering the overall walling effect from vegetation leaf mass (Yang et al. 2010) (Chap. 3).

Positive relationships are found between SVF and WVR at all LCZs (Fig. 5.7). The relationships are significant at the 5% level at all LCZs except for LCZ4. The

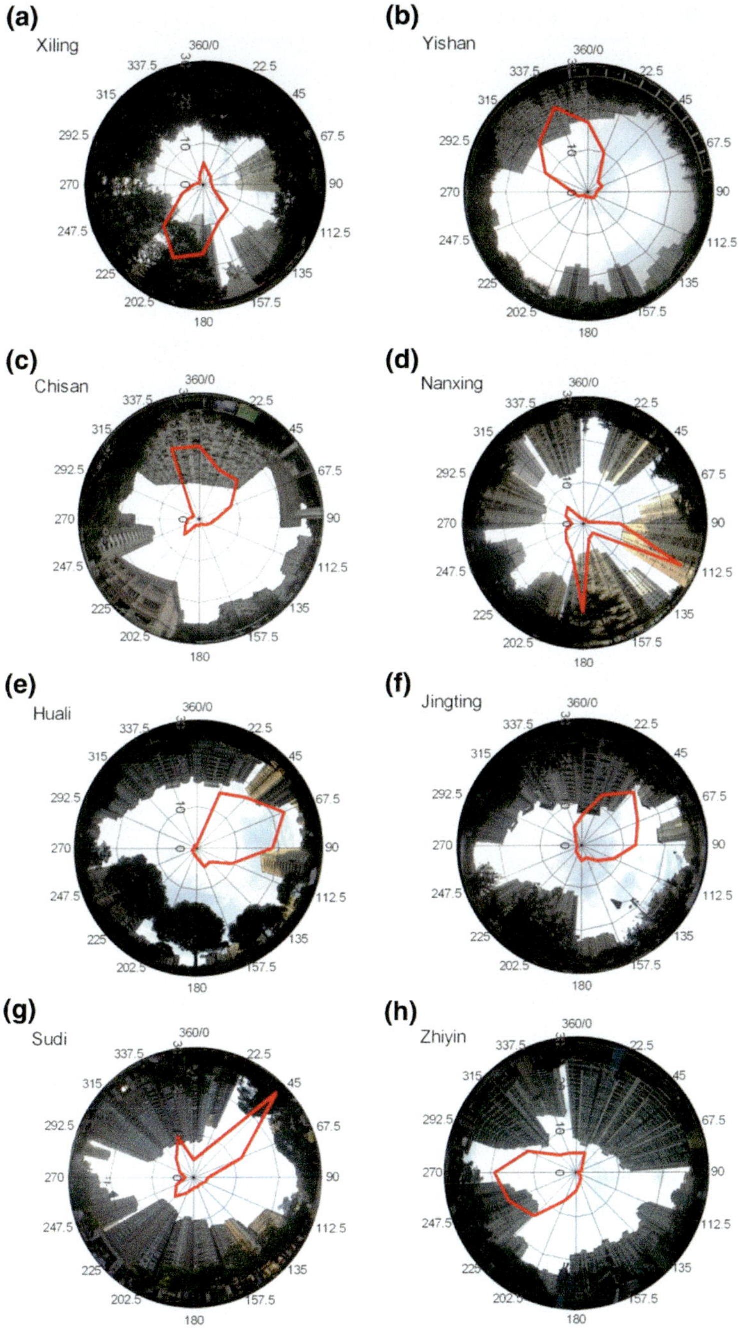

Fig. 5.6 Wind directions at the tens sites during the measurement period

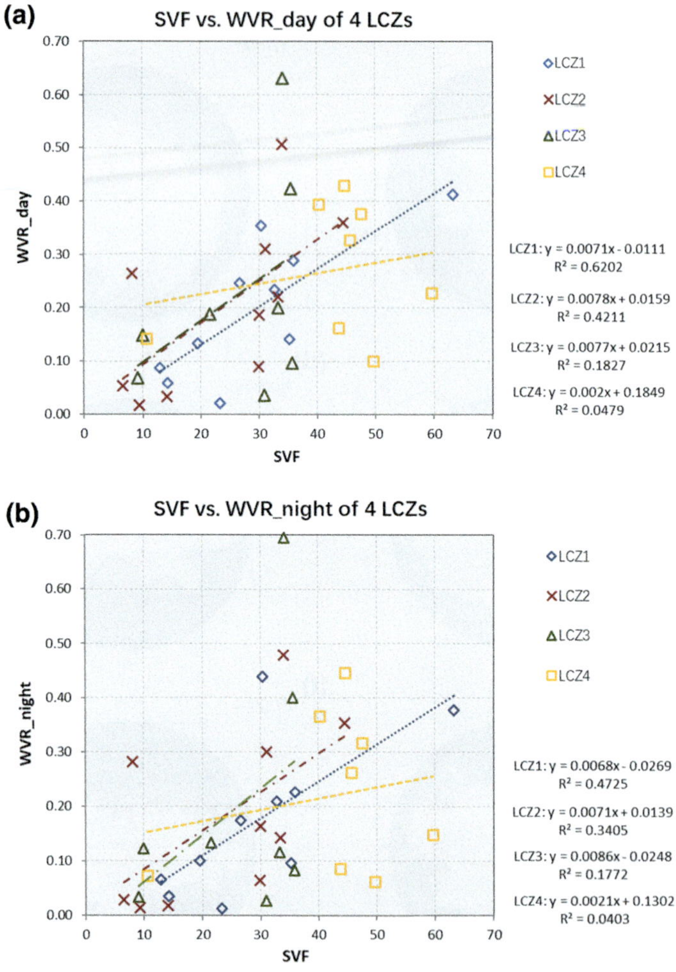

Fig. 5.7 Scatter-plot and linear-fit estimate of SVF with **a** WVR_day, and **b** WVR_night

uneven distribution of SVF values is clearly seen in this zone, with the majority cluttered within the 40–50% zone. Most of these data comes from Jingting, which features a uniform linear layout and little tree cover. This could contribute to the weak relation. If the LCZ4 data is excluded, the linear regressions of other three zones explain about 20–60% of the variability in WVR. The slopes reveal a fairly clear trend: within the range 10–50%, an increase of 10% in SVF tends to increase WVR by 7–8%, both during the day and at night.

Similarly, for TVF, significant relationships are found in all LCZs except for the LCZ4 (Fig. 5.8). The heavily uneven distribution of TVF data in this zone, largely due to the universal absence of trees (TVF ranges 0–5%) in Jingting, could influence the correlation. This kind of unsatisfactory sampling should be avoided in

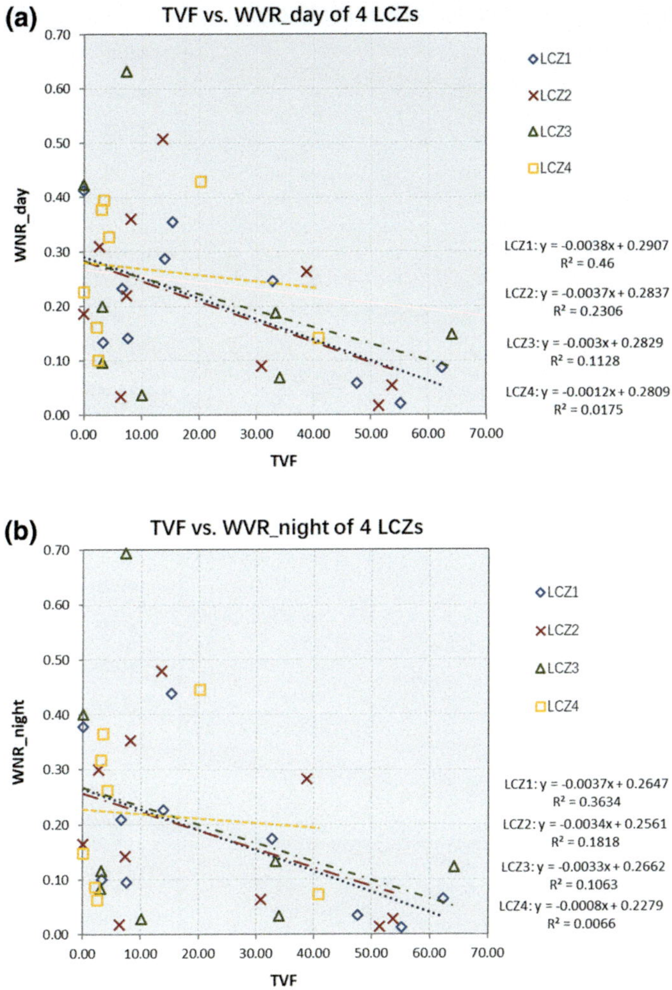

Fig. 5.8 Scatter-plot and linear-fit estimate of TVF with **a** WVR_day, and **b** WVR_night

future research work. With the LCZ4 been excluded, regression analysis reveals the explanatory power of TVF in the order of 10–45%. For all the rest three zones, the slopes indicate that, within the range 0–60%, a decrease of 10% in TVF tends to increase WVR by 3–4%, both during the day and at night.

Concerning GPR, a positive sign is found in the ER zone (Fig. 5.9). Due to the aforementioned reason, this insignificant relationship could be caused solely by chance and thus is excluded from the useful dataset. Data from the rest three zones supports consistent negative relationship of GPR with WVR, with the R-square values ranging 0.1–0.3. With GPR varying between 0 and 5, an increase of GPR of 1 tends to decrease WVR in the order of 3–7% both during the day and at night.

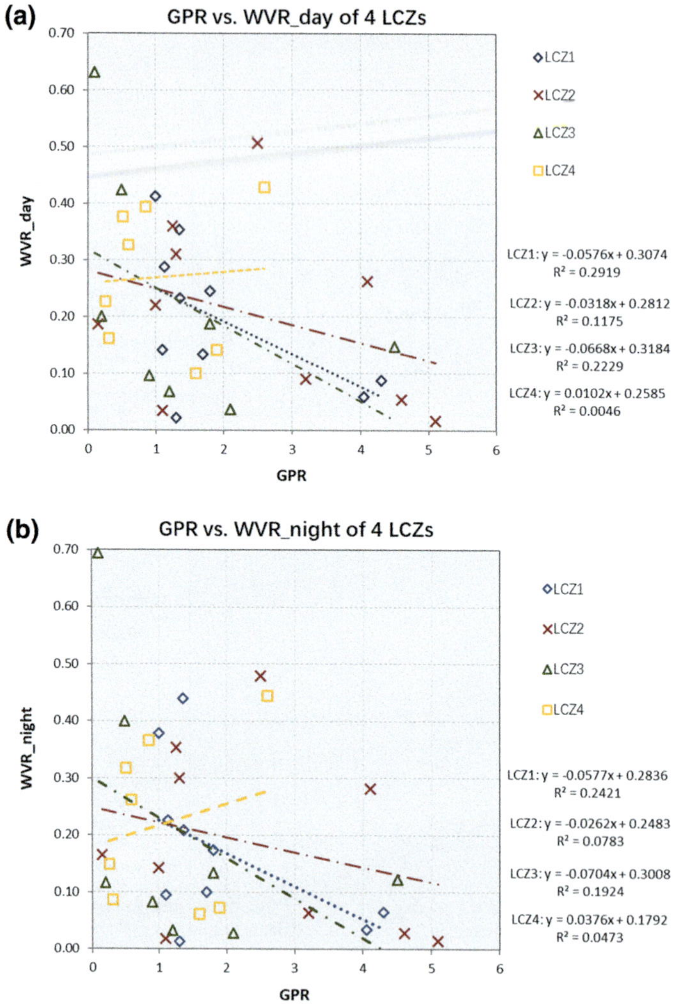

Fig. 5.9 Scatter-plot and linear-fit estimate of GPR with **a** WVR_day, and **b** WVR_night

5.1.3 Discussion

The wind effect of building layout is not clearly observed in this study. Other factors such as vegetation and density also have influenced the result. This is a drawback of the semi-experimental approach. Nevertheless, correlation analysis indicated that, overall, significant relationship exists between FAR and WVR, both at day and at night. Data from LCZ1 indicates that the least dense site (Yishan) enjoyed the highest WVR, while the densest Xiling suffered the lowest WVR. The LCZ4 data indicates that, when site layouts are similar (both are linear in this case), the site with a lower FAR (Huali) can accommodate more airflow. The LCZ3 data indicates that when FAR is similar, well-planned layout and building details in accordance with local winds can significantly increase the ventilation ratio; no clear trend regarding building volume and layout was found in the LCZ2 datasets. The result indicated that, from the ventilation viewpoint, FAR of any new high-rise residential developments in the inner city of Shanghai should be controlled below 2.5 to achieve a WVR of 0.25–0.3.

The site-mean WV, the reference WV, and the deduced WVR are all clearly lower during night hours than during the day, indicating the impact of solar radiation on the thermal-induced air flow. During the night, the suppressed mixing in the UBL leads to a more calm atmospheric condition (Oke 1987). However, night ventilation is considered one of the most important passive cooling strategies in the hot-climate area. This advertent wind profile makes effective wind environmental design even more difficult.

Most measurement points in the ten sites fell far below the comfort level (Fig. 5.3). Higher WVRs were found in sites with lower densities (e.g., Yishan and Huali) but improvements were insignificant. Tall buildings as well as trees and shrubs reduce the openness and thus the ventilation potential at the ground-level locations, the influence being indicated by the positive relationship of WVR with SVF, and negative relationships with TVF and GPR. Although tree cover could lower air temperature, above a certain threshold (i.e., 30% in this case study) it can significantly reduce WVR.

The degree of openness, quantified by SVF, can partly explain the ventilation potential at the pedestrian level. When there is advection in the UBL, SVF is related to the breathability of a pedestrian location, as previously analyzed. Under a weak wind environment, thermal-induced air movement can be a major contribution to ventilation. As such, SVF also indicates the degree of solar radiation heating, and thus the degree of buoyancy-driven airflow. SVF could therefore be used as indicators of solar incidence as well as stack ventilation potential. The correlation of SVF with WVR, the derived correlation coefficiency and explaining power in this study is based on a limited datasets from field measurement. More datasets from other sites of Shanghai and from other seasons should be acquired to test the model.

Although universally comfortable urban spaces seem not practically achievable, urban designer should not be too pessimistic, if the practical objective is to create a diverse physical environment that offer plenty of choices, or "perceived control" to users in response to various activities and situations (Steemers and Steane 2004). For instance, high WVR values were measured at certain areas in Nanxing (NX4) and Sudi (SD1), although the rest of locations were poorly ventilated. NX4 is located in the central garden of the estate. Although surrounded by six tall towers that contribute to a very high FAR (5.1) (Fig. 5.2), this place enjoyed a pleasant breeze varied around 1.5 m/s throughout the measurement period. Compared to the adjoining recreational areas that are largely covered by trees and awnings, the garden can obviously be a more welcomed place after sunset. However, residents are restricted from entering this garden by the perimeter fence. In the moderate dense site Sudi (FAR = 3.2), SD1 is located at the center of the entrance plaza. The surrounding tall buildings, although oppressive, are organized in such a way that favors the southeastern winds, and the 3-story high through passage to the northeastern side helps facilitate the cross-ventilation. As a result, air flow over this open space is sufficient to improve thermal comfort (~1.5 m/s).

5.2 The Winter Period

5.2.1 Introduction

The winter wind study intends to complement the summer study, to understand the extent and critical ranges of pedestrian-level winter winds in terms of pedestrian comfort and safety in a hot-summer cold-winter climate. We studied a residential quarter with a mixed housing building typology. With a total area of 11.2 hectare, it is divided into three sites: site A (comprised of 24-story tower buildings), site B (14 to 18-floor, high-rise linear slab buildings) and site C (6-floor mid-rise linear buildings) (Fig. 5.10). The studied sites are surrounded by other residential quarters to the Eastern, Western and Southern sides, with similar housing forms (mainly mid-rise linear residential buildings mixed with a few high-rise tower or linear residential

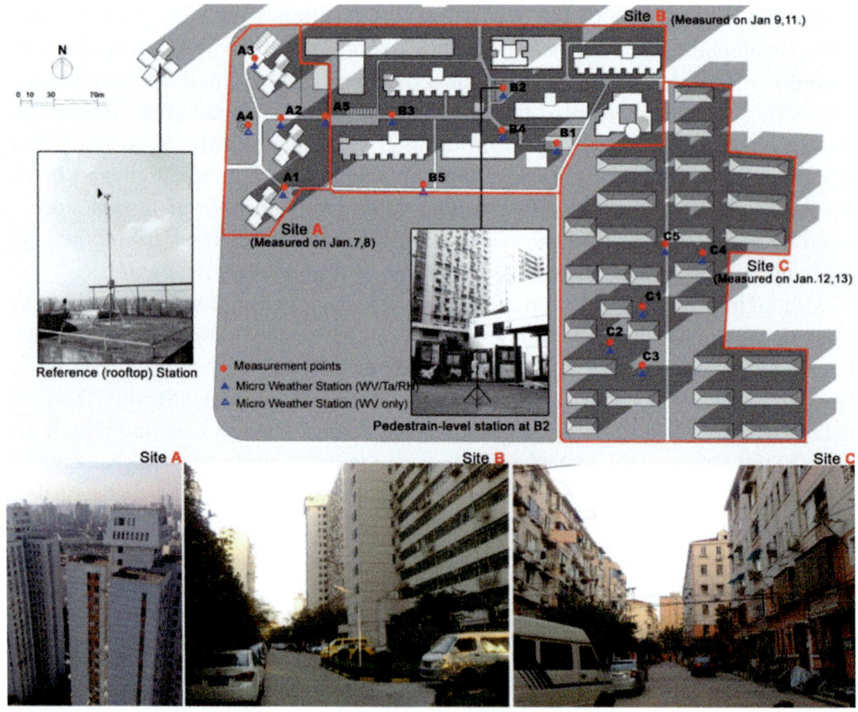

Fig. 5.10 Site plan and measurement locations

buildings). The sites are bounded by an elevated highway situated to their Northern side.

The fieldwork was carried out in January 2012 for six days. A total of fifteen locations at the pedestrian level were selected for this study and important biometeorological parameters were measured. The sampling points cover the widest possible spatial variations of the outdoor spaces. At each location, a Hobo micro weather station measured Ta, RH, and WV at a 10-second interval, from 8:30 am to 8:30 pm. The measurement lasted two consecutive days at each site.

WVR is defined as the ratio of pedestrian wind velocity to that at an upstream rooftop weather station, with sensor height $Z_S = Z_H$. The reference station was set up over the rooftop of a high-rise tower apartment building to the northwestern side of the quarter, i.e., the windward side in the winter. The WV and WD was measured with sensors mounted on top of a tall steel cast and was connected to a weather-sealed 15-channel Hobo data logger.

A questionnaire survey was conducted based on guided interview with local residents and space users, in order to gather information on subjective perception and evaluation of the respondents on the comfort and safety effect of the pedestrian-level wind environment.

The guided interview and questionnaire survey were carried out during the period of biometeorological measurement. Firstly, the demographical information, including age, gender, residence status, clothing level, physical activity level 15 min prior to being surveyed, etc., were recorded. Then three meteorological parameters (Ta, RH, and WV) were subjectively evaluated, followed by a subjective evaluation on personal acceptability toward wind environment based on the four-point scales (0: clearly acceptable; −1: just acceptable; −2: just unacceptable, and −3: clearly unacceptable) and thermal comfort evaluation based on the four-point scales (0: comfortable; −1: slightly uncomfortable; −2: uncomfortable, and −3: very uncomfortable) recommended by ISO 10551(2001). In total, 263 respondent questionnaires were collected which were considered valid for analyzes.

5.2.2 *Results*

- Wind velocity statistics

We examined the measured wind data based on accepted standard, i.e., Lawson criterion to evaluate the comfort and safety impact of the measured pedestrian wind velocity. The pedestrian wind velocity was then correlated with housing form and geometry parameters to check the possible impact of site plan, building layout, and configuration on the pedestrian wind velocity.

The local prevailing wind direction during the measurement period (8:30 am-8:30 pm) was from North and Northwest. The measurement period from 9:00 am to 8:00 pm was used for our analyzes. Therefore, for each site, the total dataset of the two-day mean wind velocity was 3960 measured at each point. The measured

Ta at the pedestrian level ranged 5–10 °C. There is one station at each site that only acquired 30-min mean wind velocity data due to equipment disorder, and thus was not used in the graphical data as shown in Fig. 5.11 and analyzed in Table 5.1, however, their mean WV data was included for statistical analysis.

The wind velocity statistics for the three sites A to C are shown in Fig. 5.11. The major wind velocity values (between the 25th and 75th percentile) at the reference rooftop station fell within approximately 3–4 m/s, 2–4 m/s, and 1–2 m/s for

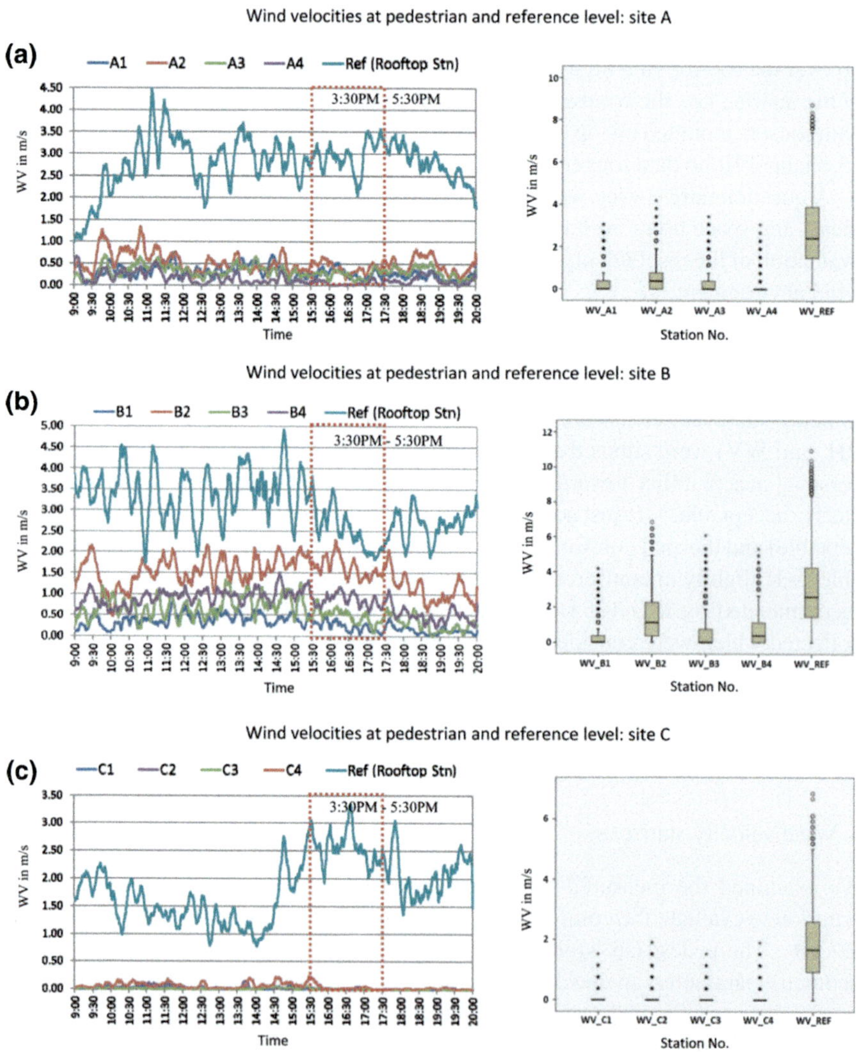

Fig. 5.11 Wind statistics: 10-min mean WV plot and WV box plots at measured locations and the reference station. **a** Wind velocities at pedestrian and reference level: site A, **b** wind velocities at pedestrian and reference level: site B, **c** wind velocities at pedestrian and reference level: site C

Table 5.1 Assessment of pedestrian's wind comfort based on Lawson criterion

	1.8 m/s		3.6 m/s		5.3 m/s	
	No. of exceedance	Ratio (%)	No. of exceedance	Ratio (%)	No. of exceedance	Ratio (%)
A1	218	3	0	0	0	0
A2	411	5	4	~0	0	0
A3	249	3	0	0	0	0
A4	64	1	0	0	0	0
B1	283	4	20	~0	0	0
B2	2822	36	527	7	66	1
B3	943	12	97	1	5	0
B4	1262	16	80	1	1	0
C1	0	0	0	0	0	0
C2	2	0	0	0	0	0
C3	0	0	0	0	0	0
C4	1	0	0	0	0	0

site A, B, and C, respectively. The pedestrian-level wind velocity data is all significantly reduced compared to the reference data. In site C, air movement at all stations appeared to be stagnated, wind velocity variations in site A and B reasonably reflected the wind effect of building. For instance, A2 and A3 are located at the Northwest side of the residential towers and exposed to the prevailing wind, whereas A1 is at the leeward side of tall buildings. The results have reflected the effect of site locations on the wind velocity pattern, in which A1 was significantly lower than A2 and A3. B2 and B4 are located along a northwest-southeast road that is connected with the main entrance of site B at the northern boundary. The wind channeling effect was clearly shown in the results, as B2 and B4 are markedly higher than B3 and B1 which are, therefore, sheltered by long-linear tall buildings at their windward side. By examining the tabulated wind velocity according to the Lawson criterion Ref (Table 5.1), the wind velocities at points A1, A2, and A3 in site A exceeded 1.8 m/s by more than 2% of the measured data, meaning unsuitable for pedestrian sitting for a long-time. In site B, all points were unsuitable for sitting for long period of time, and wind velocity at B2 exceeded 3.6 m/s by 7% of measured data, meaning unsuitable also for standing and casual walking. As for site C, a calm wind condition was recorded, and at no point the wind velocity exceeded the lowest threshold.

To effectively investigate the effect of urban form parameters (e.g., H/W, i.e., aspect ratio of street canyons) on pedestrian-level ventilation using a reference rooftop station, the wind velocity at the reference station should exceed certain threshold values, otherwise the coupling between the upper and the lower airflow can be lost, and urban materials and thermal effect can determine measured wind velocity. The threshold depends on ambient wind direction and building geometry, and can range 1.5–4 m/s for various aspect ratios (Georgakis and Santamouris 2008). Strictly speaking, the three sites in the present study cannot be classified as street canyons (i.e., L/W > 20). However, to minimize the effect of thermal buoyancy flow, a reasonably high reference wind velocity range is still preferred. In addition, the three sites should share a common range of reference velocity. Therefore, the measurement period from 3:30 pm to 5:30 pm was chosen, as during this period, the mean reference wind velocities of three sites were all within 2.5–3 m/s (site A: 2.96 m/s; site B: 2.50 m/s; site C: 2.57 m/s). Therefore, the wind data measured during this period was used to calculate microscale WVR and site-scale WVR.

Site-wise, building density variables (building cover ratio and floor area ratio) have no clear relationship with mean WVR. This is not in accordance with the summer study, in which WVR is negatively related to FAR and BCR. The reasons could be that the three sites studied in winter is adjacent to each other and thus the site effects on wind mingled, and that uneven vegetation cover may affect pedestrian wind speed, for instance the dense shrubs in site A led to lowered WVR than site B despite that site A has a lower BCR.

Point-wise, the points that are located in the road intersection, small plaza and free from tree canopy shading would be best-ventilated, as indicated by higher WV, e.g., A2 (0.17 m/s) and B2 (0.66 m/s); whereas, the points surrounded and covered by dense trees or shrubs appeared to be best sheltered from winter wind, e.g., A4 (0.05 m/s) and B1 (0.09 m/s).

To understand the contribution of housing form to the mean wind velocity, a multivariant regression was carried out. The dependent variable was 10-min mean wind velocity measured at pedestrian-level stations. The selected independent variables were reference wind velocity measured at the rooftop station (WV_{ref}), Sky view factor (SVF) and two categorical variables LA1 and LA2. LA1 and LA2 were categorical (dummy) variables to represent the impact of site building layout and configuration. $LA1 = 1$ for all datasets from site A, $LA2 = 1$ for all datasets from site B, and $LA1 = LA2 = 0$ for all datasets from site C (Yang et al. 2011). The yielded regression is shown in Eq. (5.1). The model R value is 0.72. It explains more than a half of the variation in wind velocity.

$$WV = 0.157 + 0.089\,WV_{ref} + 0.223\,LA1 + 0.776\,LA2 - 0.011\,SVF (R = 0.72) \tag{5.1}$$

From the equation, the reference wind velocity is positively related to the pedestrian wind velocity, and the housing form has a strong impact: the pedestrian wind velocity in an interspersed housing form would be averagely 0.22 m/s higher than a low-rise linear housing form, and a high-rise linear housing form would be 0.78 m/s higher than a low-rise one. From the comparison among individual points, the degree of enclosure would have an impact on ventilation rate, but correlation analysis between WVR and SVF would support a negative coefficient. The significant positive correlation of SVF with WVR derived from our summer measurement had indicated that, in hot and calm summer days, thermal-induced air movement can be a major contribution to ventilation. As such, SVF also indicates the degree of shortwave radiation heating, and thus the degree of buoyancy-driven airflow rate. In winter days, however, solar radiation is low and thermal buoyancy effect is week, wind pressure can be the major driving force. Speed of air flow at pedestrian level would be increased when section area of the air path is reduced. This is why wind velocity amplification is often found around gaps between tall buildings. This could explain the different effect of SVF on summer and winter WVR. Clearly, more empirical data and numerical study are needed before any solid conclusion can be made.

- Questionnaire survey

All measured points in site A have largely acceptable wind environment (Fig. 5.12a and b). The worst case is A3, where 26% of the respondents voted unacceptable wind environment (−2 and −3). This is because A3 is located at the northwestern corner of the estate and is highly exposed to the prevailing wind. There is one respondent voted for clearly unacceptable (−3) at A2 (4%) and A3 (3%). The vote at A2 was given by a 79-years old male resident that is physically not very well. He also voted very cold (−3) for thermal perception and uncomfortable (−2) for thermal comfort. A5 had a 100% of just acceptable votes, probably due to the small sample size (4). As for thermal comfort votes, more than a half (53%) of the respondents at A3 voted slightly uncomfortable (−1) or worse, and about one fourth of the respondents at A2 and A4 voted uncomfortable (−2). A2 to A4 are all lack of shelter from the northwesterly winds, whereas A1 and A5 are better sheltered and thus more comfortable.

The wind and comfort perception seems more diverse at site B (Fig. 5.12c and d), regarding wind acceptability, less than 5% at B1 and B5 chose unacceptable (−2 and −3), while nearly a half of respondents at B4 chose unacceptable. Higher wind velocity was measured at B2 than B4 (Fig. 5.11b), but it is considered more acceptable at B2 than B4, with less than 30% voted unacceptable wind environment. The sensor location could be an influential factor. B4 is located to the southeast of B2, on a major road of the housing quarter. The interview was carried out on the road, but the wind sensor had to be placed on the adjacent green patch to make way for vehicle traffic. The surrounding shrubs and small trees could have reduced the measured wind velocity. Concerning thermal comfort, B4 is considered uncomfortable or worse, voted −2 and −3 by more than 30% of the respondents, followed by B2 (nearly 10%). B2 and B4 are aligned with a NW-SE oriented wind corridor, whereas B1 and B5 are well-sheltered by tall buildings to their northern side.

As shown in Fig. 5.12e and f, two respondents (50%) from C3 considered the wind environment just unacceptable (−2), and the same two people considered the thermal environment uncomfortable or worse (−2 and −3). The number of valid responded questionnaires from site C is significantly smaller than that of site A and B. Considering the small sample size, the deviation at C3 should therefore be treated with caution. Besides at C3, other points were considered acceptable (0 and −1) in terms of wind, and comfortable and only slightly uncomfortable with respect to thermal comfort. This is in accordance with the measured wind data (Fig. 5.11c).

As shown in Fig. 5.13, the relationship between mean wind velocity at each point and the percentage of respondents at the point that voted clearly acceptable (0) and acceptable (0 and −1) about the wind environment (Fig. 5.13a and b) and voted

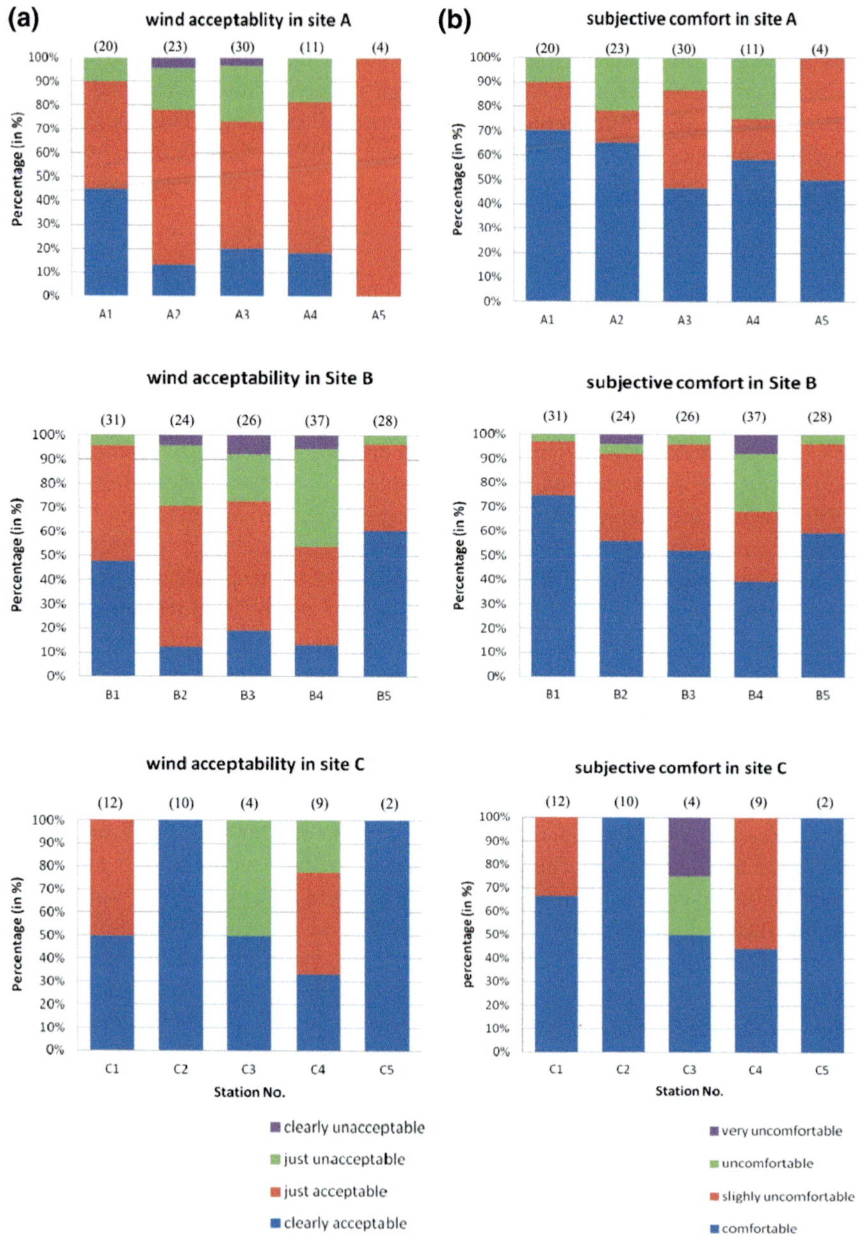

Fig. 5.12 Wind perception **a** and comfort votes **b** in three sites

Fig. 5.13 Correlation of mean WV with **a**, **b** wind perception and **c** comfort evaluation

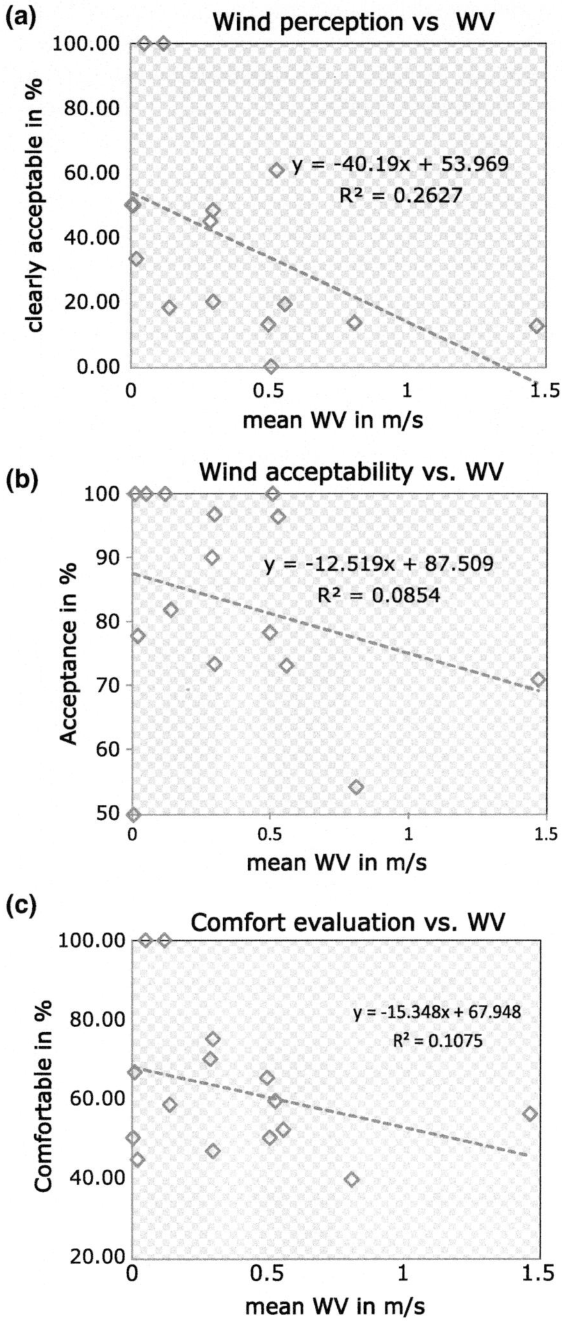

comfortable (0) about thermal comfort (Fig. 5.13c). The mean wind velocity is significantly and negatively correlated with the wind perception and comfort perception. Therefore, explaining the variation in wind perception, which was more acceptable than in thermal comfort, as indicated by the *R*-square values. The correlations are significant, but not sufficient to explain the majority of variations. This makes sense as thermal comfort is affected by other variables such as air temperature and solar radiation . The reason for the low *R*-square for wind acceptability can be that the respondents based their judgments on a short-term wind velocity experience (past 15 min) but wind velocity was highly variable and the variation is not reflected in the mean wind velocity. Also the small sample size may affect the linear-fit outcome. More datasets would need to be collected in future work, to test this preliminary result.

The overlay of site-mean wind velocity on the wind and comfort perception distribution (Fig. 5.14) indicate that, the rate of accepting the wind environment is negatively related with the mean wind velocity, so is the relationship between mean wind velocity and comfort perception. But the difference seems not significant. For

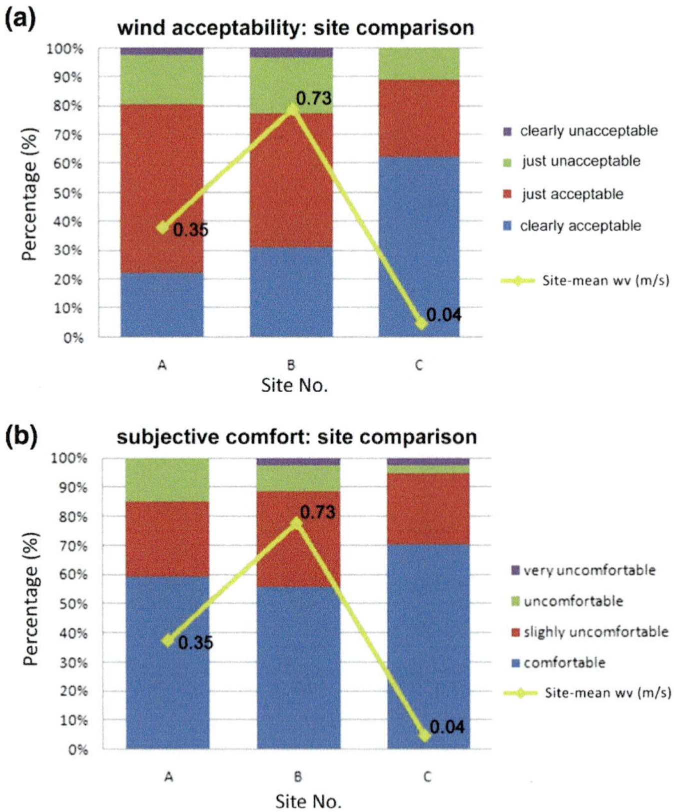

Fig. 5.14 Comparison of three sites on: **a** wind perception, and **b** comfort perception

instance, site B suffered highest wind velocity and scored lowest in both subjective wind and comfort tests, but respondents from site B that voted acceptable wind are 77%, only about 10% lower than that of site C (88%). In terms of overall comfort level, about 56% of respondents from site B voted for overall comfort, about 15% lower than that of site C. If we consider the slightly uncomfortable rating in an outdoor setting is still in the adaptable comfort range, then the difference between site B and C is even smaller: site B scores 87% which is lower than site C by only 7%.

5.2.3 Discussion

Various site layouts demonstrate different wind sheltering performance, indicating a significant impact of building form and housing typology on outdoor ventilation. The site with high-rise, long-linear slab buildings had the highest wind velocity ratio within its major outdoor spaces. The other linear layout with mid-rise linear housing form was best sheltered from the winter wind. However, according to the accepted industry criteria, even the windiest points within site B are still considered safe for passing-by pedestrians. The difference in subjective wind and comfort perception between site B and site C was not large. In fact, there are still more than three quarters of respondents from site B who considered the wind acceptable. Can we, therefore, assume that the wintertime wind shelter for high-density high-rise housing quarters in Shanghai is not critical compared with summertime ventilation requirement, and that the site planning and design should focus mainly on enhancing ventilation and consider wind shelter as secondary?

More study may be needed to answer this question. If the future work supports the above assumption, then the mid-rise linear layout would be the worst planning model in terms of site ventilation. The aspect ratio (height to width of the street canyon) of this layout, determined by the insolation right law of Shanghai housing design code is 1:1. This would be the urban canyon configuration where the skimming flow dominates; and thus, wind shelter would be enhanced but street ventilation would be restrained. This housing typology is mostly built during the 1950–70s and is still the dominant one by floor area in Shanghai. To improve the area, the housing layout could be renovated to introduce more winds into its inner space, e.g., add some tall buildings within old mid-rise housing clusters, and remove some buildings to break the long-linear canyon, to form vertical and horizontal wind corridors.

5.3 Conclusion

- In recorded summer period, an overall weak wind condition was observed. Thermal buoyancy is a major driving factor under such calm condition. This thermal airflow mechanism results in higher WV during the day than at night, but this increment is far from enough for thermal comfort, due to high air temperature. The comfort zone could be reached from early evening till nighttime (after 6 pm), when ambient temperature falls considerably. It is, therefore, important to pay more attention to the nighttime usage of outdoor spaces at the design stage, e.g., lighting design and safety facility design should be given special concern.

- At the site level, the building floor area ratio (FAR) and building coverage ratio (BCR) appears to be negatively related with WVR, indicating the wind effect of building density. However, the model coefficients are subject to examination by further work, ideally by wind tunnel study.
- At the station level, three variables, sky view factor (SVF), tree view factor (TVF), and green plot ratio (GPR) are developed to quantify the degree of enclosure at a pedestrian location. The three variables are significantly correlated with WVR as shown by three of the four LCZ data. SVF demonstrates the highest explaining power on WVR. SVF has the potential to indicate the degree of enclosure (or openness) to ambient airflow, and to shortwave radiation heating, and thus, the degree of buoyancy-driven air movement in a hot and calm environment.
- Creating diverse environment help maintain outdoor comfort under such severe thermal condition, in that it contributes to meeting the requirement of "perceived control." In practice, outdoor spaces varied in openness could be systematically organized, that offers plenty of choices in response to various users and activities both spatially and temporally. SVF has the potential as the indicator to map the solar and wind incidence, so as to inform designers and aid the climate-responsive design process.
- The measured winter WVR was highest in high-rise long-linear building layout (site B), and was lowest in mid-rise linear building layout (site C). The interspersed high-rise building layout (site A) was in the middle, but the WVR measured at site A could be lowered by the dense vegetation nearby some weather stations.
- The measured pedestrian wind velocity is largely mild. All measurement points at site B were found unsuitable for sitting around for long period and one point was uncomfortable for casual walking. Although site B has higher WVR and is windier than at C, the percentage of people that felt comfortable and slightly uncomfortable (87%) was only 7% less than at C (94%). For pedestrian spaces located in a winter wind corridor, the winter wind problem can be alleviated by proper landscape design.
- Summer wind channeling versus winter sheltering: In the hot-summer cold-winter (HSCW) building climate zone, for a given building volume, high-rise and mid-density housing form can be better than mid-rise and high-density one. The former can be windier in winter but most areas are still within comfort zone; furthermore, the windy areas can be moderated by proper landscape design, which would introduce more airflow to pedestrian level in summer, enhancing wind channeling and thus enable pedestrian comfort.

References

Blocken B, Carmeliet J (2004) Pedestrian wind environment around buildings: literature review and practical examples. J Therm Envelope Build Sci 28:107–159

Bottema M (1999) Towards rules of thumb for wind comfort and air quality. Atmos Environ 33:4009–4017

Georgakis C, Santamouris M (2008) On the estimation of wind speed in urban canyons for ventilation purposes—Part 1: coupling between the undisturbed wind speed and the canyon wind. Build Environ 43:1404–1410

Givoni B, Noguchi M, Saaroni H, Pochter O, Yaacov Y, Feller N, Becker S (2003) Outdoor comfort research issues. Energy Build 35:77–86

ISO 10551 (2001) Ergonomics of the thermal environment–assessment of the influence of the thermal environment using subjective judgement scales. International Organization for Standardization, Geneva

Ng E (2008) Policies and technical guidelines for urban planning of high-density cities—air ventilation assessment (AVA) of Hong Kong. Build Environ 44:1478–1488

Oke T, Mills G, Christen A, Voogt J (2017) Urban climates, Cambridge University Press, Cambridge

Oke TR (1987) The boundary layer climates, 2nd edn. Methuen, London and New York

Steemers K, Steane MA (eds) (2004) Environmental diversity in architecture. Spon Press, New York

Yang F, Lau SSY, Qian F (2010) Summertime heat island intensities in three high-rise housing quarters in inner-city Shanghai China: building layout, density and greenery. Build Environ 45:115–134

Yang F, Lau SSY, Qian F (2011) Urban design to lower summertime outdoor temperatures: an empirical study on high-rise housing in Shanghai. Build Environ 46:769–785

Yang F, Qian F, Lau SSY (2013) Urban form and density as indicators for summertime outdoor ventilation potential: a case study on high-rise housing in shanghai. Build Environ 70: 122–137. https://doi.org/10.1016/j.buildenv.2013.08.019

Yang F (2015) Investigating wintertime pedestrian wind environment and user perception in dense residential neighbourhood in a city of hot-summer cold-winter climate zone China. Indoor Built Environ 26 (3): 392–408. https://doi.org/10.1177/1420326x15620257

Chapter 6
Urban Thermal Radiant Environment and Heat Stress

Abstract *Study intent* This chapter discusses the impact of high-density urban form on the urban thermal environment from a human perspective: the heat stress. In contrast to the air temperature aspect that is normally investigated, the thermal radiant environment of urban settings is examined and the effective indicator of the mean radiant temperature (T_{mrt}) is used to characterize the urban thermal radiant environment and assess outdoor thermal comfort and heat stress. Two different urban settings with different building geometry and vegetation cover in downtown Shanghai were used as case study sites. A typical heat wave day in 2013 was selected to investigate the daytime radiant heat stress intensity. A GIS-based numerical simulation approach is used and the Solar and Longwave Environmental Irradiance Geometry (SOLWEIG) model was employed to investigate the spatial variation of T_{mrt}. Spatial analysis modules were developed and the Radiant Heat Stress Intensity index was defined. *Result and discussion* The study reveals that in Shanghai, under heat waves, the heat stress induced by the thermal radiant environment is quite severe, with T_{mrt} commonly well above 60 °C in daytime, and intra-urban T_{mrt} differences are largely influenced by building density and height, street orientation, and vegetation. Open paved spaces and space near sunlit walls are the places that have the highest T_{mrt}. In conclusion, the present study shows that the spatial variation of T_{mrt} can be used to identify thermally vulnerable areas and hotspots in complex urban environment and provide implications for urban design toward the mitigation of heat stress in high-density cities.

6.1 Introduction

One of the most profound impacts of high-density urban settings on the climatic condition at locale is the altering of the thermal environment. On the one hand, tall buildings create deep street canyons where dark shadows can be found; on the other hand, the buildings serve as bulk source of heat emission due to their large amount of operating energy and high-albedo surface material for ground and wall. We all have

The contents in this chapter are published in: Chen et al. (2016).

F. Yang and L. Chen, *High-Rise Urban Form and Microclimate*,
The Urban Book Series, https://doi.org/10.1007/978-981-15-1714-3_6

the unpleasant experience of walking through the sunlit side of a building in a hot summer afternoon, and feeling the walls, ground, and shadings available drastically changes our sensations. Indeed, heat stress in cities is becoming increasingly severe due to the unprecedented urbanization and global warming paces. Outdoor thermal environment in cities significantly affects citizen's behavior and lifestyle (Chen and Ng 2012). With respect to people's comfort, health, and well-being, daytime heat stress will cause health risks such as fatigue, morbidity, or even mortality in extreme weather conditions such as heat waves (Hajat et al. 2010; Gabriel and Endlicher 2011; Goggins et al. 2012). Besides the high air temperature in city centers which has been commonly observed and widely studied (Arnfield 2003; Stewart 2011), the thermal radiant aspect of the urban environment has drawn much attention only in recent years with the progress in remote sensing technologies and in situ monitoring techniques. In fact, convective fluxes of sensible heat and latent heat are more significant in directing human energy balance and thermoregulation as compared to conductive heat, especially in strong solar exposure conditions. Therefore, the thermal radiant environment plays a more important role in affecting outdoor thermal comfort in hot summer. There has been an increasing number of research studies aiming at mitigating urban heat island and reducing heat stress in cities from the thermal radiant perspective (Takebayashi and Moriyama 2007; Tan et al. 2016; Wang et al. 2016).

The mean radiant temperature (T_{mrt}) is the most important biometeorological parameter that characterizes the effect of a thermal radiant environment on human thermal comfort (Fanger 1972). Thorsson et al. (2014) have shown that daily maximum T_{mrt} can be used as an effective tool to access the level of heat stress as indicated by mortality. In this chapter, we will investigate the spatial variation of T_{mrt} as affected by different urban geometry and vegetation in high-density cities and its implication for heat stress during heat waves. Two different urban settings in Shanghai's downtown area were selected as the study sites. A GIS-based numerical simulation approach was taken and the Solar and Longwave Environmental Irradiance Geometry (SOLWEIG) model [31, 36] was used to simulate T_{mrt} pattern for a real summer hot day. To evaluate heat stress intensity from the thermal radiant aspect, we define the index of Radiant Heat Stress Intensity (RHSI) to count for both the intensity and the duration of the heat stress. Spatial analysis of the spatial variations of T_{mrt} was conducted, based on which heat stress areas were identified and urban design implications for mitigating urban heat stress were proposed.

6.1.1 Study Area Description

Shanghai (30°40′N ~ 31°53′N, 120°51′E ~ 122°12′E), the biggest city in China, is located on the alluvial terrace of the Yangtze River delta with an average elevation of 4 m above sea level. It has a northern subtropical monsoon climate, with a mean annual T_a of 17.2 °C, and a monthly maximum and minimum mean T_a of 30.2 °C in

July and 1.9 °C in January for the last 30 years, respectively. The city's administrative boundaries cover a total terrestrial area of 6340.5 km^2 excluding estuary waters. The city has a total population of 24.15 million, including both permanent and non-permanent residents. As one of the most rapidly urbanizing cities in China, Shanghai is also suffering increasingly severe heat stress. In 2013, the city experienced its hottest summer in the last 140 years since weather data was recorded. Eight consecutive days of T_a above 38 °C and 25 days of T_a above 35 °C were recorded in July. The highest official T_a of 40.6 °C was also recorded by the city's benchmark observatory. During the hot spell, hospital admissions generally increased by 30%, and at least 10 people died from heatstroke (BBC 2013; Shanghai Daily 2013). In such cases, identifying urban hot spots and areas vulnerable to heat stress is particularly important for implementing protection measures.

The Lujiazui (LJZ) central business district is located in the Pudong New District on the eastern bank of the Huangpu River just across the Bund. The area has a mixture of commercial, business, and high-end residential land uses. There are a total number of 45 super high-rise buildings taller than 200 m in LJZ, and the tallest building, Shanghai Center, is 632 m high. The area is also characterized by abundant urban open spaces and greenery including parks and roadside trees. In comparison, the old residential neighborhood of Xiaonanmen (XNM) is located in the Huangpu District on the western bank of the Huangpu River. The area is densely packed with old residential buildings normally 1–4 story high. Although there are a few tall buildings built more recently, none of them is taller than 120 m. Because of the limited space, vegetation such as trees and shrubs is extremely scarce in XNM, and in some narrow streets, there is literally no urban greenery. The area is also dominated by the regular E–W and N–S street orientations. These two areas represent two different urban settings of the high-dense city center of Shanghai. Figure 6.1 shows the locations of LJZ and XNM in Shanghai and snapshots of their typical street scenarios. Two

Fig. 6.1 Left: Locations of downtown center in Shanghai along the Huangpu River. Middle: Satellite image showing the study sites of LJZ (indicated by the red square on the top) and XNM (indicated by the red square at the bottom). Right: Typical street scenarios in LJZ (up) and XNM (down)

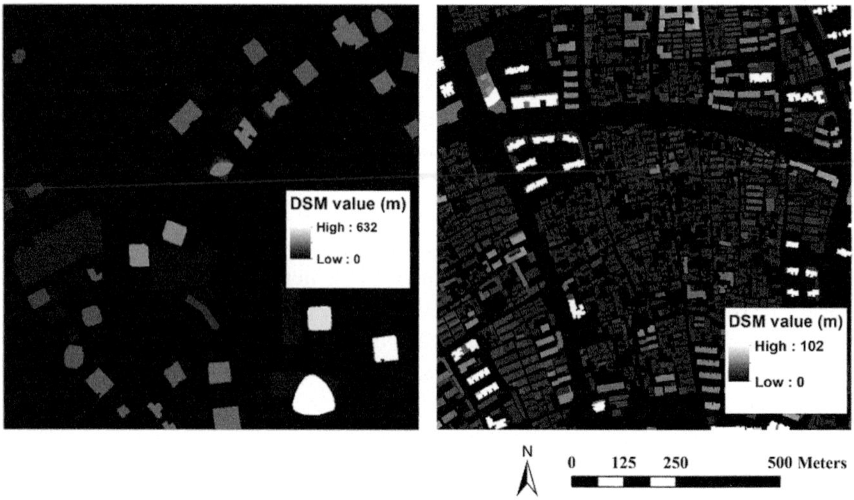

Fig. 6.2 DEM maps of LJZ (left) and XNM (right)

Table 6.1 Comparison of statistics of morphological information of LJZ and XNM

Site	Total building floor area (10^6 m^2)	Built-up ratio (%)	Average building height (m)	Average SVF (sky view factor)/standard deviation
LJZ	4.82	18.0	107.1	0.65/0.17
XNM	1.46	36.5	12.9	0.53/0.21

1 by 1 km sites were selected from LJZ and XNM as case study sites. Figure 6.2 shows the digital elevation model (DEM). Table 6.1 summarizes the morphological information of the two sites.

6.1.2 A GIS-Based Numerical Simulation Approach

The Solar and Longwave Environmental Irradiance Geometry (SOLWEIG) model (Lindberg et al. 2008; Lindberg and Grimmond 2011) is a computational model that can simulate the spatial variation of T_{mrt} in a complex urban environment. SOLWEIG calculates T_{mrt} by simulating the 3-D radiant flux densities of the surrounding environment, which is based on the integral 3-D radiant environment monitoring technique proposed by Höppe (1992) and has been shown to calculate T_{mrt} accurately (Thorsson et al. 2007). The model uses a digital surface model (DSM) as primary input data and implements fast algorithms for raster calculation. In this way, it can deal with the real urban environment and calculate T_{mrt} values for a large spatial domain in a computationally efficient manner. Therefore, SOLWEIG has advantages over other computer models that either are too time-consuming such as the holistic simulating package of ENVI-met (Bruse and Fleer 1998) or can only handle the point of interest such as Rayman (Matzarakis et al. 2007). SOLWEIG has been tested in a number of cities with various urban contexts and different climates (Lindberg et al. 2008; Lindberg and Grimmond 2011; Chen et al. 2014; Lau et al. 2016). Constant improvements have been made to the model, such as including the effect of vegetation (Lindberg and Grimmond 2011) and surface material (Lindberg et al. 2016). In this study, the latest version of SOLWEIG 2015a is used, which requires meteorological data consisting of time-series of T_a, relative humidity (RH) and global solar radiation (G), as well as DSMs of building and vegetation.

6.1.3 Model Verification and Validation

To validate the SOLWEIG model in Shanghai's unique climatic context, 3-D radiant flux density monitoring campaigns were conducted on the university campus on two clear hot summer days, i.e., August 25 and 17, 2015. Two cases were tested. One is a paved open square in the west of an N–S aligned building with limited vegetation on the western side (square case); the other is an E–W oriented street canyon with much vegetation (street canyon case). The two selected sites can be used to evaluate SOLWEIG's performance in near-wall and open space cases, street canyon, and also cases when vegetation is included. One set of net radiometer (Kipp and Zonen CNR 4) was used to measure the 3-D radiant flux densities. The radiometer was rotated every 3 minutes to each of the three directions (E–W, N–S, U–D). The monitoring was conducted from 8:00 to 16:00 for each day, following standard measurement routines proposed by (Höppe 1992; Thorsson et al. 2007). SOLWEIG simulations were performed for the two days and compared with field measurement data for model validation. Figure 6.3 shows the fish-eye lens photographs taken at the measurement points and the DSMs showing the point of interest used for validation.

SOLWEIG simulations were performed for the two selected sites in downtown Shanghai. One clear and calm hot summer day in a heat wave, i.e., July 28, 2013, was selected for investigation. Meteorological data recorded by meteorological stations of the Shanghai Meteorological Bureau was used to simulate the real case of the spatial variation of T_{mrt}. There is a station located in the center of LJZ and another station located about 2 km to the west of XNM. Normally, T_a for XNM is 2–3 °C higher than LJZ in summer daytime primarily due to the high density. For the simulation day, the maximum T_a was 38.7 °C and 40.1 °C for LJZ and XNM, respectively. Figure 6.4 shows the hourly variation of major meteorological parameters for the two sites on the simulation day.

Building DSMs were generated from 3-D building database acquired from Shanghai Institute for Surveying and Mapping, with the resolution of 1 m. Vegetation DSM for LJZ was generated from airborne LiDAR data (Huang et al. 2013), which was

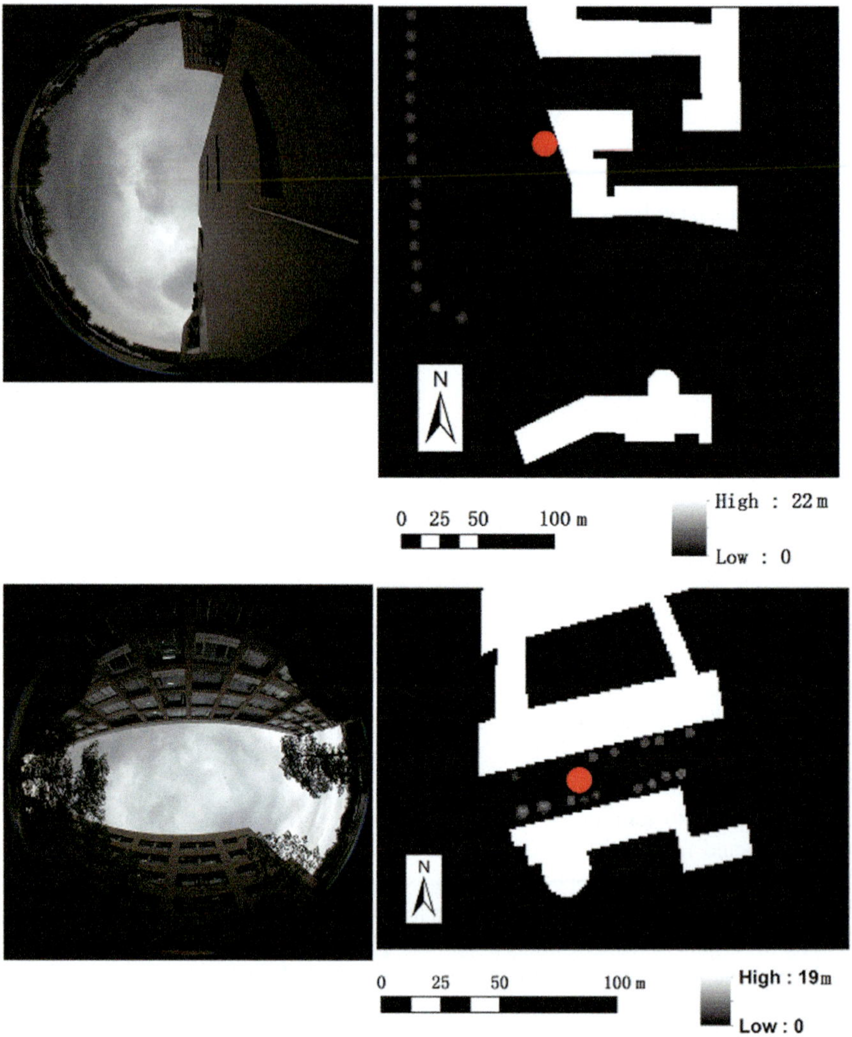

Fig. 6.3 Left: Fish-eye lens photos taken at the measurement point in the square case (up) and the street canyon case (down). Right: DSMs of two validation sites, i.e., square case (up) and street canyon case (down), the red dots indicate the location of the measurement points and points of interest in SOLWEIG

further corrected and supplemented by on-site surveys based on the profile of greenery provided by Google Map satellite images. Standard absorption coefficients for shortwave and longwave radiation were used, i.e., 0.7 and 0.97, respectively (Höppe 1992). The surface albedo for building and ground surfaces was set to 0.2. The emissivity of building walls was set to 0.9 representing a combination of concrete and glass, and the emissivity of ground surfaces was set to 0.95 for rough concrete. The

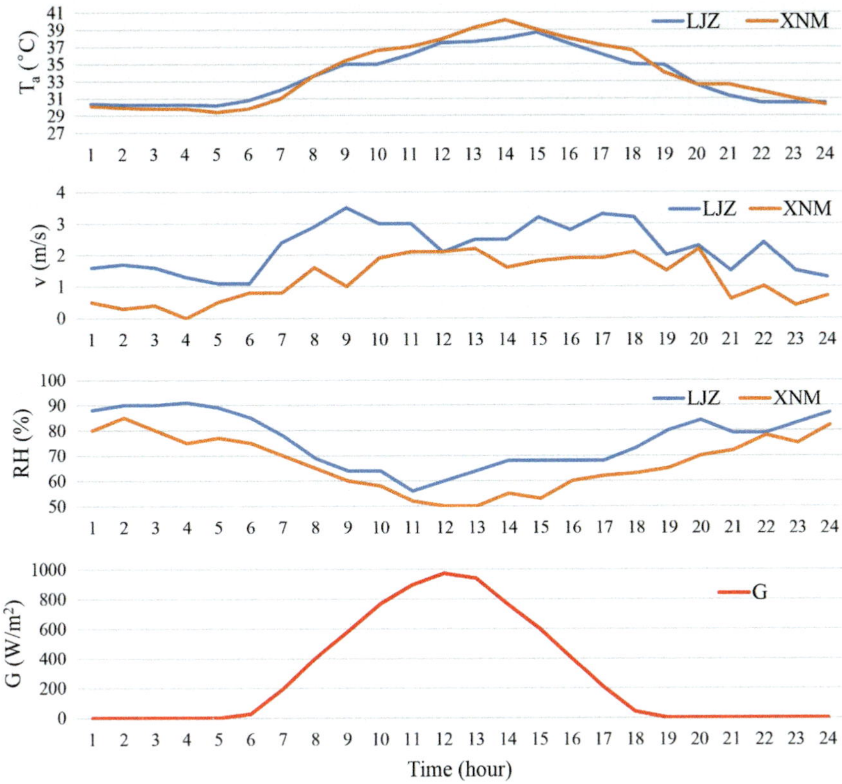

Fig. 6.4 Hourly variation of major meteorological parameters for both sites on July 28, 2013: air temperature (T_a), wind speed (v), relative humidity (RH), and global radiation (G)

transmissivity of shortwave and longwave radiation through vegetation was set to 0.05 and 0, respectively, as recommended by (Lindberg and Grimmond 2011). Base on the input DSMs, meteorological data, surface characteristics, and Shanghai's geographical location, T_{mrt} values were calculated for a standing or walking person, using the surface projection factors of 0.22 for the east, west, north, and south directions and 0.06 for the upward and downward directions (Mayer and Höppe 1987; Höppe 1992; Kántor and Unger 2011). The simulations were conducted for each hour from 9:00 to 16:00.

6.2 Results

6.2.1 Model Performance

Modeled shortwave and longwave radiant flux densities at 15 min interval from 09:00 to 16:00 were compared with measurement data. Downward, upward, side, and total shortwave and longwave radiant flux densities were compared, denoted as K_{down}, K_{up}, K_{side}, K_{total} for shortwave and L_{down}, L_{up}, L_{side}, L_{total} for longwave, which is also the settings used by (Lindberg et al. 2008; Lau et al. 2016). Figures 6.5 and 6.6 show the correlations between modeled and measured shortwave and longwave radiant flux densities for the square case and street canyon case, respectively. As the figures show, the performance of the SOLWEIG model is very good. For the square case, when little vegetation is present, SOLWEIG can explain 96.2% of the variance in K_{total} (RMSE = 130.7 W/m^2) and 95.6% of the variance in L_{total} (RMSE = 41.4 W/m^2). For the street canyon case, when a lot of vegetation is present, SOLWEIG can explain 94.7% of the variance in K_{total} (RMSE = 142.2 W/m^2) and 85.3% of the variance in L_{total} (RMSE = 43.0 W/m^2).

More careful examinations were given to different components of K_{total} and L_{total} to evaluate model performance in detail. For the square case, the modeled K_{down} and K_{up} agreed with measured values very well, i.e., R^2 = 0.93, RMSE = 60.9 W/m^2 for K_{up}, and R^2 = 0.92, RMSE = 14.4 W/m^2 for K_{down}. The good correlation was expected, as the calculation of these two components was straightforward based on input radiation data and sky obstruction. For the lateral directions, noticeably, SOLWEIG underestimated K_{east}, with R^2 = 0.83 and RMSE = 31.8 W/m^2. This was especially the case when the building wall which was very close (~2 m) to the east of the measurement point got sunlit in the early afternoon, and high near-wall shortwave fluxes were observed that could be up to 70 W/m^2 higher than modeled values. This discrepancy was believed to be caused by high reflection from building surfaces. SOLWEIG also overestimated K_{west}, with R^2 = 0.81 and RMSE = 83.6 W/m^2, especially after 11:00 when the sun was visible, and the overestimation was believed to be caused by the lack of building and vegetation data in the west direction outside the campus. K_{south} was slightly overestimated, with R^2 = 0.87 and RMSE = 29.9 W/m^2, which was caused by not considering trees to the south of the site. On the other hand, K_{north} was estimated very accurately, with R^2 = 0.91 and RMSE = 7.1 W/m^2, because buildings and trees outside the site were far away, thus had little influence on the measurement point. For longwave radiation, SOLWEIG can also simulate longwave radiant fluxes very well, i.e., L_{down}: R^2 = 0.89, RMSE = 8.3 W/m^2; L_{up}: R^2 = 0.96, RMSE = 9.0 W/m^2; L_{south}: R^2 = 0.96, RMSE = 6.1 W/m^2; L_{north}: R^2 = 0.93, RMSE = 9.5 W/m^2; L_{east}: R^2 = 0.84, RMSE = 11.4 W/m^2; L_{west}: R^2 = 0.95, RMSE = 7.5 W/m^2. The biggest discrepancy was found for L_{east} (near-wall case), and SOLWEIG constantly underestimated L_{east} by over 20 W/m^2 when the wall got sunlit. This underestimation conflicts with findings reported in (Lindberg et al. 2008). Since the distance to the nearby wall was not specified in the paper,

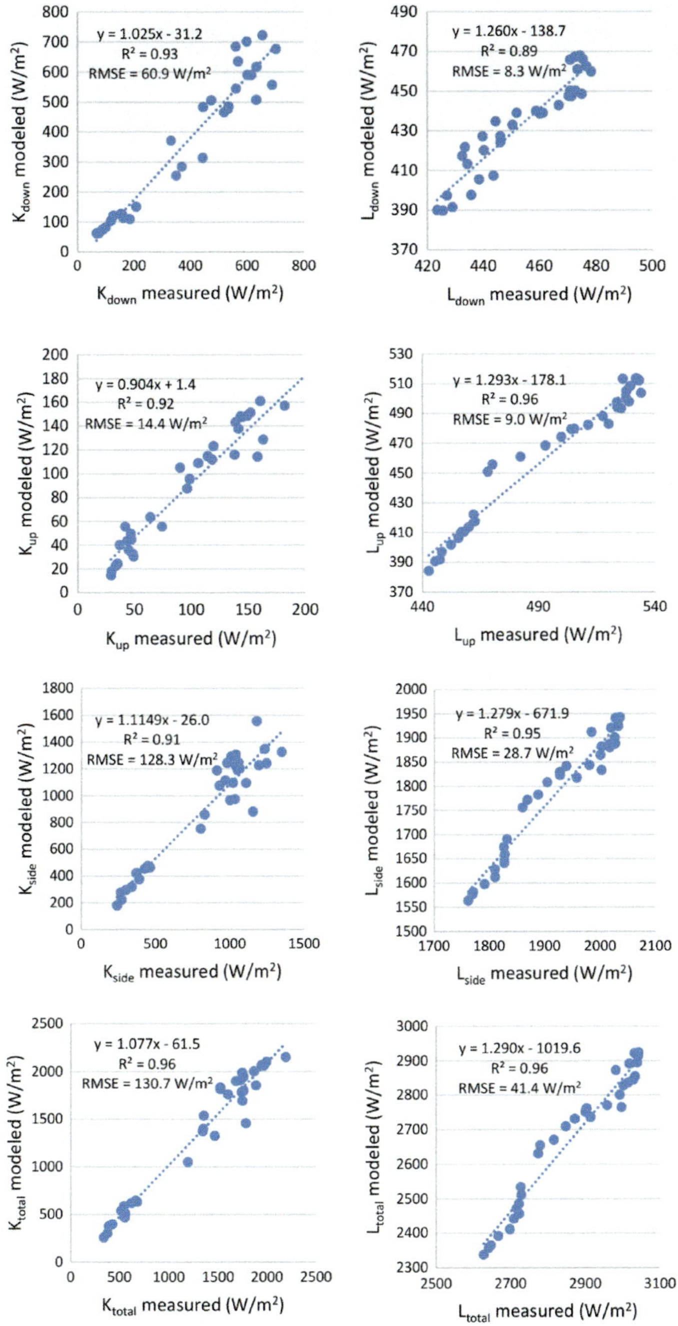

Fig. 6.5 Correlations between modeled and measured shortwave and longwave radiant flux densities at the measurement site for the square case

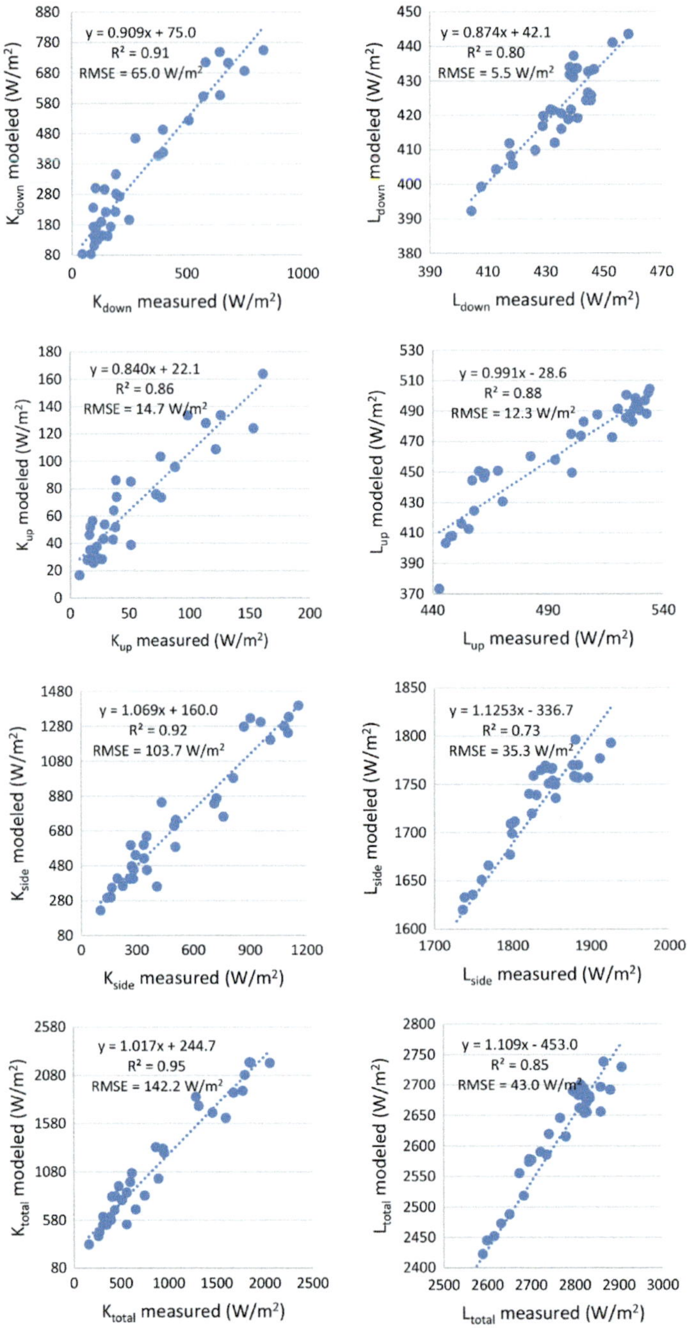

Fig. 6.6 Correlations between modeled and measured shortwave and longwave radiant flux densities at the measurement site for the street canyon case

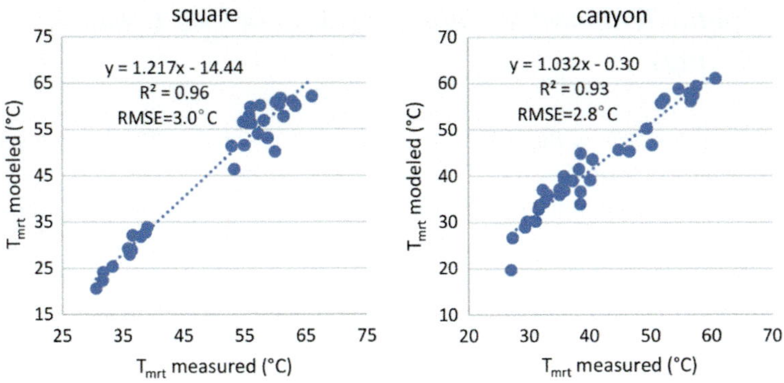

Fig. 6.7 Correlation between modeled and measured T_{mrt} at the two validation sites: (left) square case; (right) street canyon case

the cause of such disagreement could not be further examined. Nevertheless, future work should address the near-wall case in detail.

For the street canyon case, SOLWEIG simulated both shortwave and longwave radiant flux densities considerably well, but with relatively lower R^2 as compared to the square case (K_{down}: $R^2 = 0.91$, RMSE = 65.0 W/m^2; K_{up}: $R^2 = 0.86$, RMSE = 14.7 W/m^2; K_{side}: $R^2 = 0.92$, RMSE = 103.7 W/m^2; L_{down}: $R^2 = 0.88$, RMSE = 5.5 W/m^2; L_{up}: $R^2 = 0.86$, RMSE = 12.3 W/m^2; L_{side}: $R^2 = 0.79$, RMSE = 35.3 W/m^2). The reason for the lower correlation is because the irregular shape of tree trunks cannot be fully parameterized in the model. Noticeably, modeled K_{north} and K_{south} showed lowest correlations with measurement data, with $R^2 = 0.65$ and RMSE = 19.8 W/m^2 for K_{north}, as well as $R^2 = 0.71$ and RMSE = 41.8 W/m^2 for K_{south}. Because the study site is an E–W oriented street canyon, the discrepancy could be attributed to the reflection from building surfaces. In the case of K_{south}, when the building facade got sunlit at noontime, measured value could be over 40 W/m^2 higher than modeled value.

Figure 6.7 shows the comparisons between measured and modeled T_{mrt} values for a standing person based on the flux density data for the two cases. The figure reveals that modeled T_{mrt} values showed good agreement with measured values, with R^2 higher than 0.9. Noticeably, in the early afternoon with strong shortwave radiation reflected from building facades, SOLWEIG tended to underestimate T_{mrt} by around 3 °C for the near-wall case. Other than that, SOLWEIG normally overestimated T_{mrt} by around 4 °C, which could be attributed to the overestimated lateral long-wave radiant fluxes, which is consistent with findings reported in (Lindberg et al. 2008; Lau et al. 2016). The comparisons proved that with accurate input data of building morphology and vegetation cover, SOLWEIG can accurately simulate the spatial variation of T_{mrt} in various urban settings. The applicability of SOLWEIG in Shanghai's climatic context was therefore validated.

6.2.2 *Spatio-Temporal Variations of* T_{mrt} *in Two Urban Settings*

Figures 6.8 and 6.9 show the hourly spatial variations of T_{mrt} for the two selected sites in downtown Shanghai, i.e., LJZ and XNM on July 28, 2013, from 11:00 to 15:00, representing the hottest time of a day. Table 6.2 summarizes the statistics of T_{mrt} values in the two sites. It is shown that on a hot day, daytime T_{mrt} in Shanghai's urban center is very high in general, normally well above 40 °C even in shaded areas. During the study period, both sites showed similar temporal pattern of T_{mrt} changes: at 11:00, the solar radiation is already very high (~890 W/m^2), and because of the high T_a (above 36 °C), T_{mrt} can be over 70 °C in sunlit areas and 40 °C in shaded areas. As T_a keeps increasing and solar radiation is constantly above 800 W/m^2, T_{mrt} reaches its maximum value at 14:00. For LJZ, maximum T_{mrt} of 84.4 °C was found at the paved open square near waterfront in the northwest of the site, primarily due to the extremely high shortwave radiation reaching and reflected from the ground. In the meantime, minimum T_{mrt} of 50.8 °C was found at the deep shadow casted by high buildings and trees.

For XNM, because the local T_a can be up to 2 °C higher than LJZ, T_{mrt} can be considerably higher than 80 °C in some extreme cases. Maximum T_{mrt} of 90.0 °C was found near the south-facing wall in the wide E–W road in the center of the site, which is the result of intense shortwave and longwave radiations both from the ground and the wall. In fact, in the sunlit locations of this road, T_{mrt} values all exceed 87 °C, meanwhile minimum T_{mrt} of 52.1 °C was found at the western side of the shaded N–S street canyon in the densely built-up areas in the center of the site. It is noticeable that LJZ shows a much more diverse spatial distribution of T_{mrt} as compared to XNM, with standard deviation commonly above 10 °C during the study period, in contrast, XNM's standard deviation of T_{mrt} is normally around 5–6 °C. The difference is also reflected in the local scale. For example, in LJZ, for a tall building with a large open paved space in the south, the T_{mrt} values in front of the south-facing wall of the building could be over 30 °C higher than the T_{mrt} values in the deep shadow in the northwest of building. In comparison, in XNM, T_{mrt} values in shaded street canyons are normally around 15 °C lower than the nearby open space or center of wide roads, meanwhile T_{mrt} values in the northern sunlit side of E–W street canyons are normally 5–7 °C higher than T_{mrt} values in the western shaded side of nearby N–S street canyons. Nevertheless, the difference of T_{mrt} between different locations across the site can be above 35 °C during the study period for both LJZ and XNM, which is consistent with the 30 °C range as shown by Mayer and Höppe (1987).

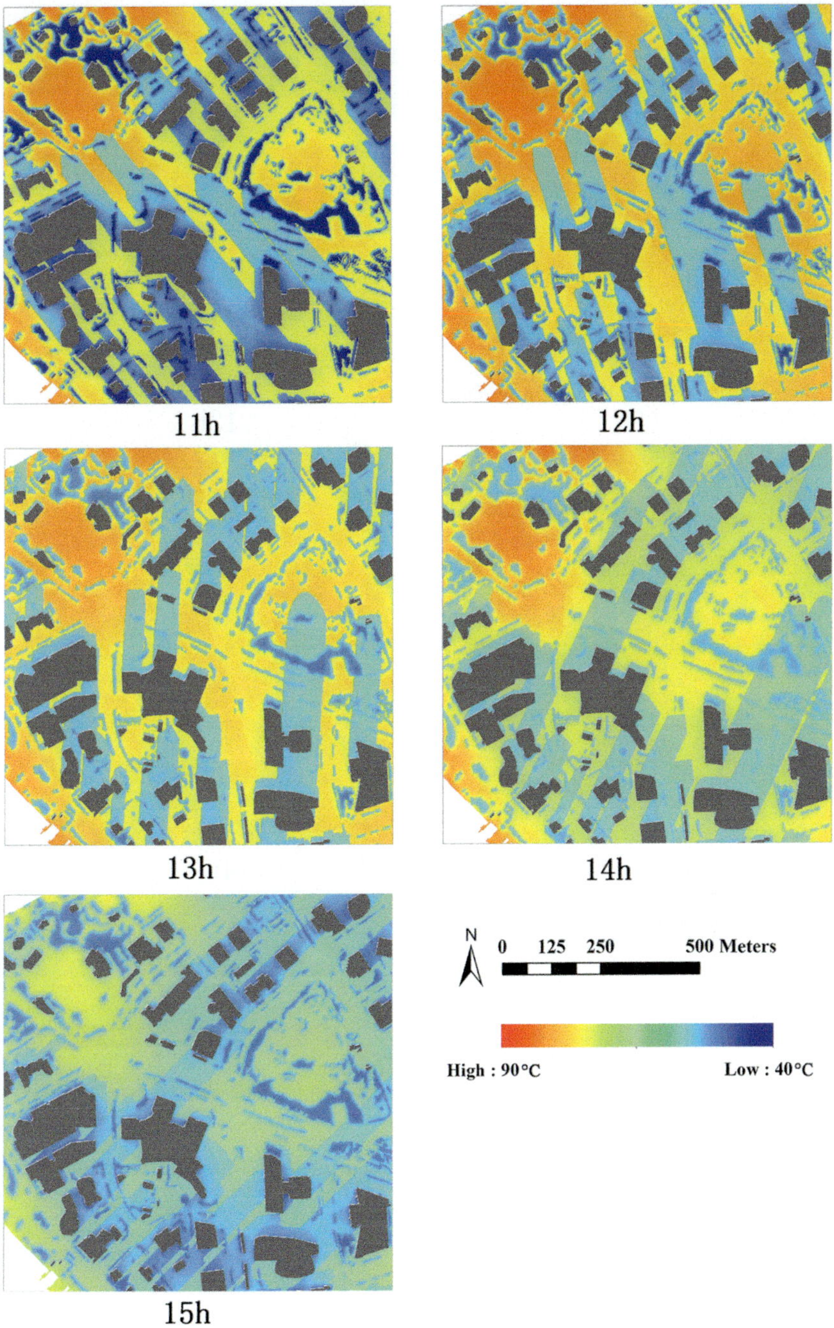

Fig. 6.8 Hourly T_{mrt} maps from 11:00 to 15:00 for LJZ

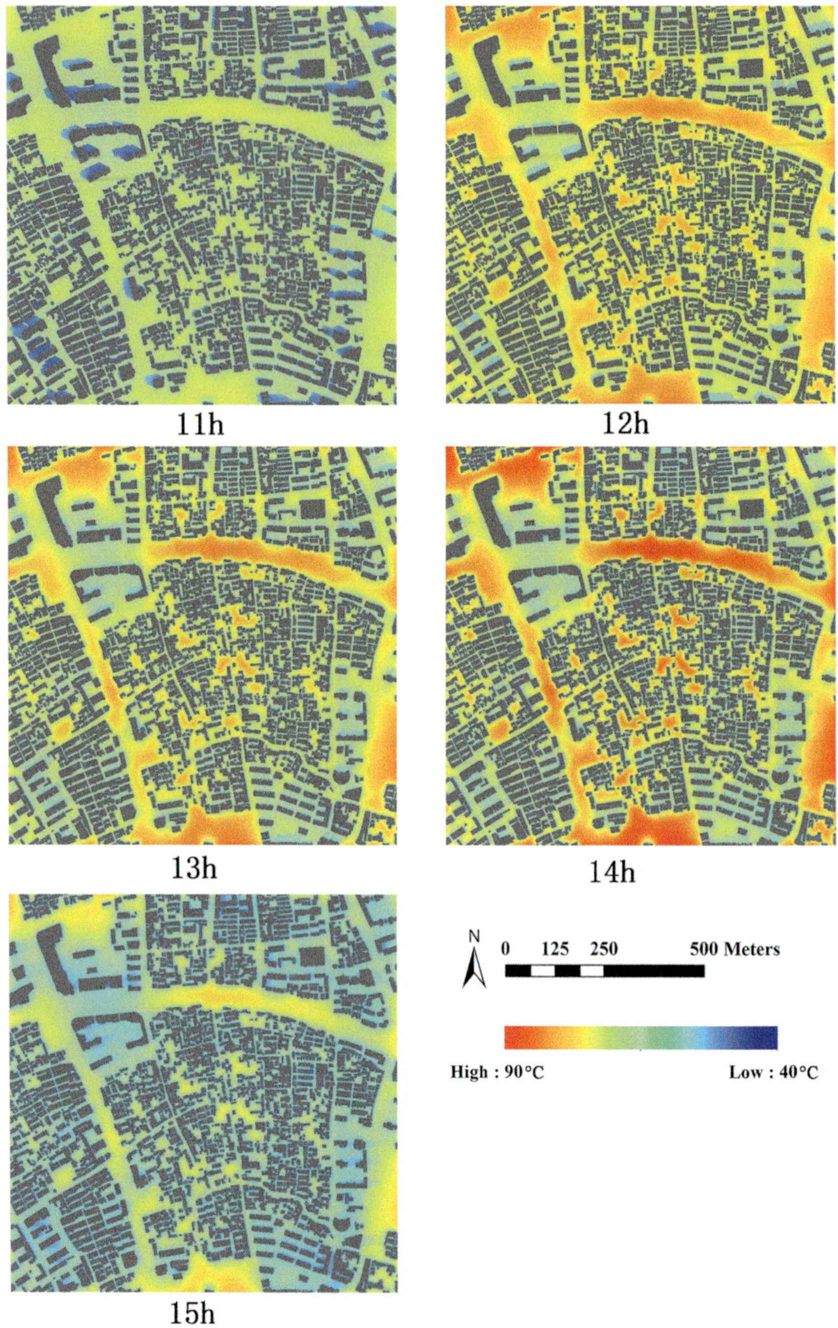

Fig. 6.9 Hourly T_{mrt} maps from 11:00 to 15:00 for XNM

Table 6.2 Statistics showing T_a, G, maximum T_{mrt} (max), minimum T_{mrt} (min), average T_{mrt} (avg), and standard deviation of T_{mrt} (std) for LJZ and XNM from 11:00 to 15:00

	LJZ						XNM					
Time	T_a (°C)	G (W/m^2)	T_{mrt} (°C)				T_a (°C)	G (W/m^2)	T_{mrt} (°C)			
			max	min	avg	std			max	min	avg	std
11:00	36.1	894.5	83.9	40.9	62.7	12.5	37.0	894.5	73.3	45.4	64.2	6.9
12:00	37.5	972.2	84.4	46. 6	66.7	12.0	37.9	972.2	83.7	52.1	71.2	5.3
13:00	37.6	939.9	82.5	48.6	67.4	10.7	39.2	939.9	85.1	50.8	68. 8	6.7
14:00	38.0	763.2	84.4	50.8	66.5	13.1	40.1	763.2	90.0	52.1	70.4	6.3
15:00	38.7	601.3	74.8	46.3	59.3	6.2	39.0	601.3	79.0	46.4	61. 9	5.6

6.2.3 *Effect of Urban Geometry on Heat Stress: GIS-Based Spatial Analysis*

In order to focus on the spatial variations of T_{mrt} and examine the intra-urban differences of T_{mrt} in the two different urban settings in Shanghai, average T_{mrt} values from 11:00 to 15:00 were calculated. Figure 6.10 shows the average hourly T_{mrt} values from 11:00 to 15:00 in LJZ and XNM. Figure 6.11 shows the histograms of the T_{mrt} values in two maps. It is found that in LJZ, maximum average T_{mrt} of 82.0 °C occurred at the center of the paved open square near waterfront in the northwest of the site, which was in constant sun exposure from 11:00 to 15:00 and received

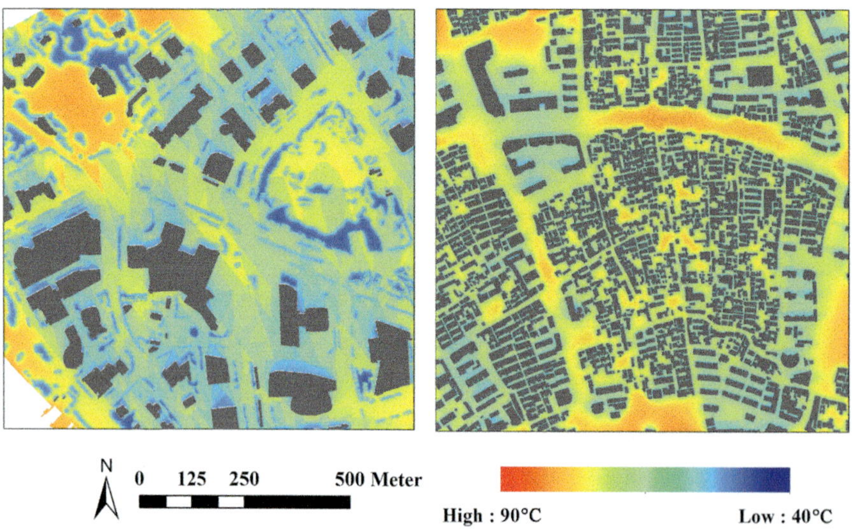

Fig. 6.10 Average hourly T_{mrt} maps from 11:00 to 15:00 of LJZ (left) and XNM (right)

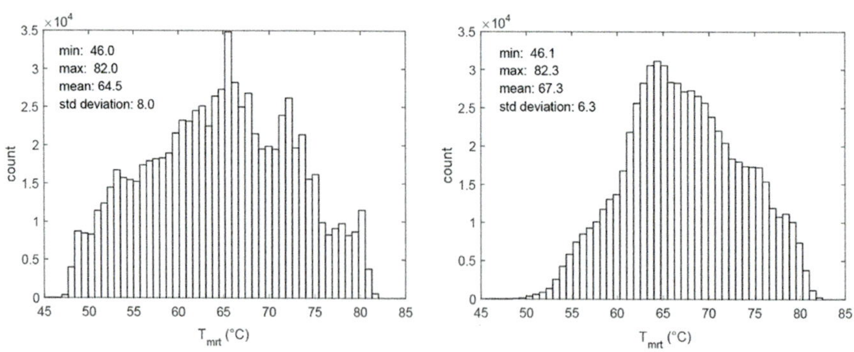

Fig. 6.11 Histograms of T_{mrt} values in two maps of LJZ (left) and XNM (right)

extremely high longwave and shortwave radiation from the sky and also reflected from the ground. In other unobstructed spaces, average T_{mrt} values were normally below 80 °C, because of the shadow casted by nearby buildings or trees. Minimum average T_{mrt} of 46.0 °C was found in the deep shadow casted by tall buildings and trees. The whole site's T_{mrt} values follow a normal distribution, with mean T_{mrt} value equal to 64.5 °C and standard deviation equal to 7.9 °C. In comparison, the T_{mrt} values across XNM follow a similar normal distribution, but the curve is much steeper, with mean T_{mrt} value equal to 67.3 °C and standard deviation equal to 6.3 °C. The distribution patterns suggest that the spatial variation of T_{mrt} in XNM is less diverse as compared to LJZ, with larger proportion of higher T_{mrt} values. In fact, high T_{mrt} values above 70 °C were commonly found in street canyons.

To further examine the effect of urban geometry on the intra-urban differences of T_{mrt}, which includes building height and density, and street orientations and width, GIS-based spatial analysis modules were developed to distinguish urban spaces of different types, such as street canyon spaces, near-building spaces and open spaces, and extract local T_{mrt} values for analysis. For LJZ, three types of urban spaces were identified, including north side of buildings which are within 10 m from a building's north-facing walls, south side of buildings which are within 10 m from a building's north-facing walls, and open spaces which are the places more than 100 m away from any buildings. Figure 6.12 shows the statistical analysis result of the T_{mrt} values in the three types of urban spaces in LJZ. It shows that the majority of spaces in northern side of buildings have relatively low T_{mrt} values around 50–55 °C due to the shading by the buildings. In contrast, the T_{mrt} values in spaces in south side of buildings are much higher, with mean value equal to 63.6 °C, not only because of the incoming strong shortwave and longwave radiations, but also because of the intense radiations reflected from the sunlit south-facing walls. The spatial variation of T_{mrt} in open spaces far away from buildings shows a more diverse pattern, due to the highly varied shadow conditions created by buildings and trees. On the other hand, the mean value above 66 °C also suggests that open spaces are generally very hot in summer daytime.

For XNM, two types of urban spaces were identified, i.e., E–W street canyons and N–S street canyons. In practice, a tolerance angle of 30° was allowed so street canyons not exactly E–W or N–S oriented were also included. Figure 6.13 shows the statistical analysis result of T_{mrt} values in E–W and N–S street canyons in XNM. It shows that T_{mrt} values in E–W street canyons are considerably higher than N–S street canyons, with a difference of 1.8 °C in the mean value. T_{mrt} is higher in E–W canyons mainly because the northern part of the canyons is normally sunlit and also close to sunlit south-facing walls. In contrast, N–S street canyons especially narrow canyons are usually in shadow casted by buildings in the west. The result agrees with the findings in (Lau et al. 2016). Another noticeable feature is that T_{mrt} values in E–W and N–S street canyons both follow normal distributions with majority values concentrating in the 60–70 °C range, reflecting the homogenously heat-dominant thermal radiant environment in street canyons.

The above discussions demonstrate that in different urban settings, the urban geometry plays an important role in affecting the local thermal radiant environment,

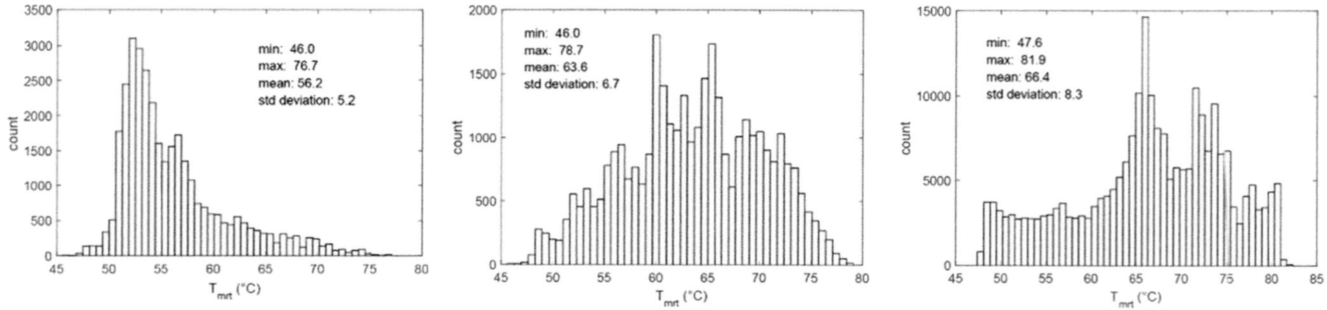

Fig. 6.12 Histograms of the statistics of T_{mrt} values in three types of urban spaces in LJZ: building north (left), building south (middle), and open space (right)

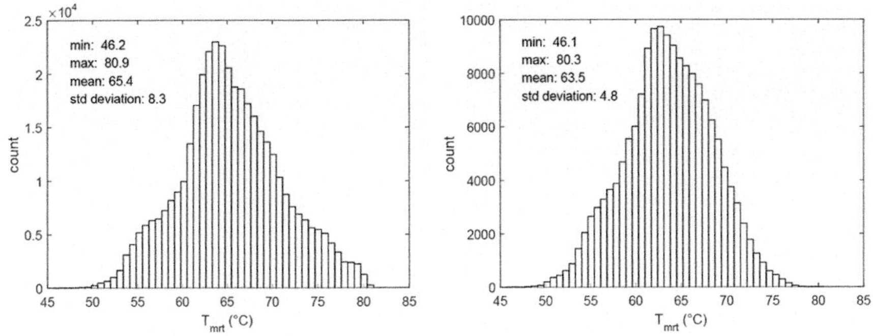

Fig. 6.13 Histograms of the statistics of T_{mrt} values in E–W (left) and N–S (right) street canyons in XNM

as indicated by the spatial variation of T_{mrt}. It shows that, in urban environment with relatively open spaces with tall buildings such as LJZ, unobstructed open spaces have extremely high T_{mrt}. Near-building spaces in the southern side of buildings also have very high T_{mrt} which is influenced by both the ground surface and the south-facing walls. In contrast, northern and western spaces of tall buildings tend to have lower T_{mrt} especially when trees are presented. In denser urban environment with low and moderate building height such as XNM, E–W street canyons especially the spaces in front of south-facing walls have very high T_{mrt}. Closed courtyards also have high T_{mrt} due to the lack of shading and also reflected shortwave radiation and longwave radiation from ground and wall surfaces, which is consistent with findings in the city of Gothenburg (Lau et al. 2015). On the other hand, narrow N–S street canyons have lower T_{mrt} because of the abundant shading by surrounding buildings.

6.2.4 *The Radiant Heat Stress Intensity (RHSI): Implications for Heat Stress and Countermeasures*

Thorsson et al. (2014) found that the T_{mrt} value of 59.4 °C is an important threshold value in predicting heat-related risk of mortality. In this study, the T_{mrt} value of 60 °C was considered as critical threshold value indicating the occurrence of heat stress in Shanghai. To consider both the intensity and duration of such heat stress induced by solar radiation, the index of Radiant Heat Stress Intensity (RHSI) was introduced, which is defined as the accumulative hourly difference of T_{mrt} above 60 °C for all the sun hours in a day. It was found that the RHSI range is 0–148.6 °C h for LJZ and 0–137.6 °C h for XNM. The wide range suggests that even for spaces where heat stress occurred, large variation in terms of the lasting impact of heat stress can still be observed. In this sense, the RHSI range was classified into five heat stress levels: 0–10 indicating low heat stress, 10–20 indicating moderate heat stress, 20–50 indicating high heat stress, 50–100 indicating very high heat stress, and 100 and

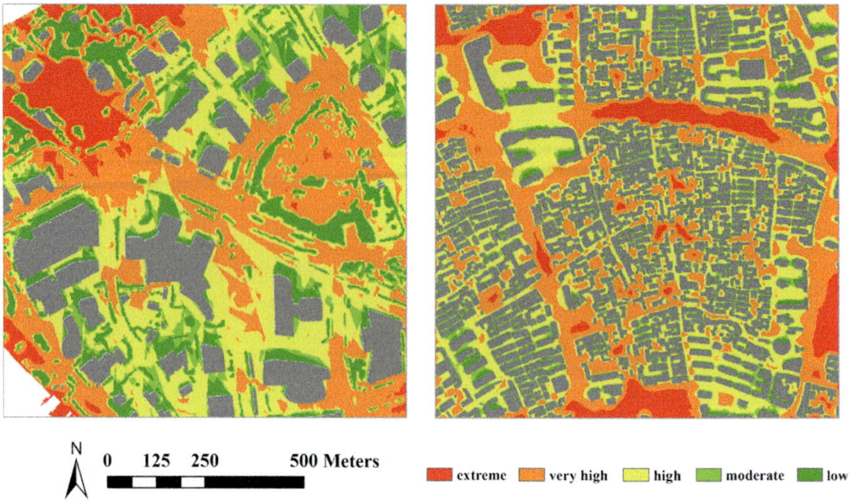

Fig. 6.14 RHSI class map on the simulation day for LJZ (left) and XNM (right)

above indicating heat extremes. Although there are other ways to select threshold values and define heat stress levels, the used method presents one logical and easy-to-interpret way to parameterize the spatial variations of the severity of heat stress in urban spaces. Figure 6.14 shows the maps of RHSI classes in LJZ and XNM on the selected hot day, i.e., July 28, 2013.

The figure shows that paved open spaces in city center have the most severe heat stress, such as the large open squares in LJZ and the wide and unobstructed E–W roads in XNM. Such heat extremes are a combining effect of intense incoming and reflected shortwave radiation, longwave radiation from ground surface, and longer sun exposure. For example, some locations in the center of the large square in the northwest of LJZ could be under sunshine for more than 11 h a day, which induces extremely high heat stress. It was also observed that open spaces in the southern side of buildings can also have very high heat stress, which is affected by the reflected shortwave radiation from sunlit south-facing walls. On the other hand, low and moderate heat stress was found in the northern side of tall buildings where abundant shading was casted and under tree canopies in urban green spaces. As one of the most prominent financial centers and tourist attractions in China, LJZ has a lot of shopping centers, commercial office buildings, and sightseeing spots located at different places. The very high heat stress in waterfront, squares, and also spaces around buildings can be very unpleasant for pedestrians on hot summer days. Since the space between buildings is especially important for the local prosperity and vitality (Gehl 2011), effective shading devices, such as trees or kiosks, should be provided for pedestrian spaces and major walkways in order to mitigate the local heat stress, which can benefit hundreds of thousands of office commuters and tourists every day in hot summer and therefore encourage the use of outdoor spaces. The large spacing between buildings should also be reduced to create denser urban fabrics and bring more shade to the spaces between buildings.

It was also found that in the denser urban environment of XNM, heat extremes also occurred in closed courtyards surrounded by densely packed low-rise buildings, which is primarily due to lack of shading, longer sun exposure, and also the effect of surrounding walls. Heat stress was normally more severe in E–W street canyons, with a vast majority having high heat stress, especially on the northern side of the streets which received less shading. Since XNM is dominated by old residential neighborhoods with a lot of senior citizens living around, these places with high heat stress are especially vulnerable to extreme climatic events such as heat waves. Due to the limited outdoor space in XNM, planting trees on the street may be hardly feasible. Therefore, sunshades along walkways protecting pedestrians from direct sun exposure are the most effective measures to mitigate heat stress and improve pedestrians' thermal comfort. In contrast to E–W street canyons, N–S street canyons exhibited moderate heat stress or even low heat stress in some very narrow streets, suggesting that street orientations and building alignments could be appropriately controlled to mitigate heat stress in densely built-up areas, such as the design strategies proposed by (Ali-Toudert and Mayer 2007; Andrade and Alcoforado 2008; Thorsson et al. 2011). On the other hand, the use of shelters in narrow street canyons and modification of street orientations can also block the local ventilation. Therefore, countermeasures to mitigate heat stress should be evaluated against the local microclimatic context in order to achieve the best effect in terms of improving outdoor thermal comfort.

6.3 Conclusion

In this chapter, a specific aspect of the urban thermal environment which is most related to heat stress the thermal radiant environment was introduced. The biometeorological index of T_{mrt} was used and the RHSI index was defined to parameterize the impact of heat stress induced by solar radiation by considering both the intensity and duration of severe radiant heat. The intra-urban differences of T_{mrt} in Shanghai on a hot summer day were studied through a GIS-based simulation approach using the SOLWEIG model. Two different urban settings representing different urban fabrics in Shanghai's downtown center were selected as case studies. Heat stress areas were identified and classified based on the spatial variations of T_{mrt}. The results show that large open squares with little vegetation have the most severe heat stress, which is a result of long exposure to direct shortwave radiation. Spaces near sunlit south-facing walls and E–W street canyons also have very high heat stress, as a combing result of reflected shortwave radiation and longwave radiation from surrounding walls. Simple shading devices protecting pedestrians from direct sun exposure is the most effective way to immediately mitigate the local heat stress, and there are other design-level countermeasures such as reducing the spacing between buildings to introduce more shadow on walkways, re-designing street orientations to have more N–S street canyons, and also changing surface materials to reduce emitted longwave radiation from building walls. It is noteworthy that these strategies should be evaluated against other climatic aspects such as ventilation and implemented in accordance with other design criteria in order to achieve a holistically optimized effect to improve pedestrian thermal comfort.

Employing the GIS techniques, the study also reveals that spatial analysis based on continuous values over large spatial domains and consequent zonal understandings are substantially important for studying the intra-urban variation patterns in complex urban environment. These analyses and understandings are only possible when spatial information is present. Such spatial information is of great importance in design and planning practices, and with the support of computers and GIS platforms, the information can be captured, stored, analyzed, managed, and visualized at the most local level. This kind of spatial decision-support process can spur the discussions on sustainable design and planning for cities by providing cause–effect consequences of potential strategies and actions.

References

Ali-Toudert F, Mayer FH (2007) Effects of asymmetry, galleries, overhanging facades and vegetation on thermal comfort in urban street canyons. Sol Energy 81:742–754

Andrade H, Alcoforado M-J (2008) Microclimatic variation of thermal comfort in a district of Lisbon (Telheiras) at night. Theoret Appl Climatol 92(3–4):225–237

Arnfield AJ (2003) Review: two decades of urban climate research: a review of turbulence, exchanges of energy and water, and the urban heat island. Int J Climatol 23:1–26

BBC (2013) China issues heat alert as 'hottest july' hits Shanghai. http://www.bbc.com/news/world-asia-china-23499175. Retrieved from 2 June 2016

Bruse M, Fleer H (1998) Simulating surface-plant-air interactions inside urban environments with a three-dimensional numerical model. Environ Model Softw 3:373–384

Chen L, Ng E (2012) Outdoor thermal comfort and outdoor activities: a review of research in the past decade. Cities 29:118–125. https://doi.org/10.1016/j.cities.2011.08.006

Chen Y-C, Lin T-P, Matzarakis A (2014) Comparison of mean radiant temperature from field experiment and modelling: a case study in Freiburg, Germany. Theoret Appl Climatol 118:535–551

Chen L, Yu B, Yang F, Mayer H (2016) Intra-urban differences of mean radiant temperature in different urban settings in Shanghai and implications for heat stress under heat waves: a GIS-based approach. Energy Build 130:829–842. https://doi.org/10.1016/j.enbuild.2016.09.014

Fanger PO (1972) Thermal comfort. Analysis and application in environment engineering. McGraw Hill Book Company, New York

Gabriel KMA, Endlicher WR (2011) Urban and rural mortality rates during heat waves in Berlin and Brandenburg, Germany. Environ Pollut 159:2044–2050

Gehl J (2011) Life between buildings: using public space, 6th edn. Island Press

Goggins WB, Chan EYY, Ng E, Ren C, Chen L (2012) Effect modification of the association between short-term meteorological factors and mortality by urban heat islands in Hong Kong. Plos One 7:e38551

Höppe P (1992) A new procedure to determine the mean radiant temperature outdoors (in German). Wetter und Leben 44:147–151

Hajat S, O'Connor M, Kosatsky T (2010) Health effects of hot weather: from awareness of risk factors to effective health protection. Lancet 375:856–863

Huang Y, Yu B, Zhou J, Hu C, Tan W, Hu Z, Wu J (2013) Toward automatic estimation of urban green volume using airborne LiDAR data and high resolution remote sensing images. Frontiers of Earth Science 7(1):43–54

Kántor N, Unger J (2011) The most problematic variable in the course of human-biometeorological comfort assessment—the mean radiant temperature. Cent Eur J Geosci 3(1):90–110

Lau KK-L, Lindberg F, Rayner D, Thorsson S (2015) The effect of urban geometry on mean radiant temperature under future climate change: a study of three European cities. Int J Biometeorol 59(7):799–814

Lau KK-L, Ren C, Ho J, Ng E (2016) Numerical modelling of mean radiant temperature in high-density sub-tropical urban environment. Energy Build 114:80–86

Lindberg F, Grimmond CSB (2011) The influence of vegetation and building morphology on shadow patterns and mean radiant temperatures in urban areas: model development and evaluation. Theoret Appl Climatol 105:311–323. https://doi.org/10.1007/s00704-010-0382-8

Lindberg F, Holmer B, Thorsson S (2008) SOLWEIG 1.0—modelling spatial variations of 3D radiant fluxes and mean radiant temperature in complex urban settings. Int J Biometeorol 52(7):697–713

Lindberg F, Onomura S, Grimmond CSB (2016) Influence of ground surface characteristics on the mean radiant temperature in urban areas. Int J Biometeorol. https://doi.org/10.1007/s00484-016-1135-x (in press)

Matzarakis A, Rutz F, Mayer H (2007) Modelling radiation fluxes in simple and complex environments—application of the RayMan model. Int J Biometeorol 51(4):323–334

Mayer H, Höppe P (1987) Thermal comfort of man in different urban environments. Theoret Appl Climatol 38:43–49

Shanghai Daily (2013) An official 40.6 °C makes it another record day for city. http://www.shanghaidaily.com/Metro/society/An-official-406C-makes-it-another-record-day-for-city/shdaily.shtml. Retrieved from 2 June 2016

Stewart I (2011) A systematic review and scientific critique of methodology in modern urban heat island literature. Int J Climatol 31:200–217

Takebayashi H, Moriyama M (2007) Surface heat budget on green roof and high reflection roof for mitigation of urban heat island. Build Environ 42:2971–2979

Tan Z, Lau KK-L, Ng E (2016) Urban tree design approaches for mitigating daytime urban heat island effects in a high-density urban environment. Energy Build 114:265–274

Thorsson S, Lindberg F, Björklund J, Holmer B, Rayner D (2011) Potential changes in outdoor thermal comfort conditions in Gothenburg, Sweden due to climate change: the influence of urban geometry. Int J Climatol 31:324–335

Thorsson S, Lindberg F, Holmer B, Eliasson I (2007) Different methods for estimating the mean radiant temperature in an outdoor urban setting. Int J Climatol 27(14):1983–1993

Thorsson S, Rocklöv J, Konarska J, Lindberg F, Holmer B, Dousset B, Rayne D (2014) Mean radiant temperature—a predictor of heat related mortality. Urb Clim 10(2):332–345

Wang Y, Berardi U, Akbari H (2016) Comparing the effects of urban heat island mitigation strategies for Toronto, Canada. Energy Build 114:2–19

Chapter 7
Cooling Effects of Urban Greenery at Three Scales

Abstract *Study intent* Greenery in a high-rise urban environment provides accessible amenity space and moderated microclimate for the local community. At the urban-to-local scale, urban greenery contributes to heat island mitigation; at the microscale, landscape elements moderate pedestrian-level thermal comfort; and at the building scale, nearby trees and facade greening or, vertical greening system (VGS), mediate building heat gain and improve indoor comfort and indoor energy performance without occupying valuable urban land. This chapter synthesizes a series of case-based investigations on the thermal environmental performance of urban greenery at three scales. The first case attempted to understand the local urban effects on the cooling performance of greenery under a simplified LCZ scheme. The second case compared two housing block design strategies to reduce summertime heat gain in outdoor space, i.e., high-albedo pavement and vegetation, particularly trees, through field measurement and numerical simulation. The third case experimentally studied the energy and thermal performance of a double-skin green façade system on an office building. *Results and discussion* The local urban density and land cover properties have notable effect on the microscale cooling rates of greenery. Tree canopy shading and overall foliage density, qualified by the tree view factor and green plot ratio, respectively, were capable of partly explaining the variation in air temperature reduction. To optimize cooling performance, residential greenery should be integrated into the local green network system. Regarding the two strategies, i.e., albedo and greenery, the second case study reveals that increasing greenery cover, especially tree cover, improves thermal comfort from mid-day till early night. In contrast, increasing ground surface albedo overall worsens the thermal comfort during the day with a marginal benefit at night. At the building scale, the experiment reveals that the double-skin green facades on both south and north façade create a distinctive microclimate in the cavity. The exterior wall surface temperature reduction of the southern facade reached maximally 9 °C, whereas the interior surface reduced maximally 2 °C. The resultant indoor thermal and energy improvements by the VGS were evaluated accordingly.

This chapter's content is published in: Yang et al. (2015, 2018).

F. Yang and L. Chen, *High-Rise Urban Form and Microclimate*,
The Urban Book Series, https://doi.org/10.1007/978-981-15-1714-3_7

7.1 The Local Scale

7.1.1 Introduction

Sizable urban green spaces (urban parks, forests, etc.) have been well studied in terms of the thermal environmental effect on surrounding urban areas. In contrast, small green patches such as residential greenery scattered in the densely built-up urban area are less studied. Residential green space is hardly comparable to sizable urban green spaces in terms of area, biomass, or diversity of plant species, but its aggregated size indicates an underestimated asset of the urban ecological system. As mentioned in Chap. 1, in the inner city of Shanghai, the residential green space comprises 23% of the overall 12.28 km^2 of green space (Che and Song 2001). Although structurally, city's residential greenery is essentially a network of small patches and features the highest degree of landscape fragmentation and the lowest fractal dimension, and the opportunity to optimize its environmental performance should be explored because given its aggregated size, it may play a role in combating the exacerbating inner-city UHI effect. One important aspect of such study would be, to better understand the impact on cooling performance of small green space by the geometry and fabric of surrounding built structures (Oke et al. 1989). This context-dependence of vegetation cooling is less investigated in previous studies (Bowler et al. 2010). Studying on the vegetation cooling in localized urban climate will better instruct urban and landscape design at the local scale.

This study aims to gain an empirical understanding of the microscale cooling by residential greenery in high-rise residential quarters in the inner city of Shanghai, and the extent of the impact on the cooling effect from local effect of urbanization. Grouping of the experimental sites is based on a simplified local climate zoning (LCZ) process. A setoff categorical variable is introduced to account for the urban to local effects on microclimate measured on site. The properties of individual sites, the field measurement protocol and parametrization of greenery into tree view factor (TVF), green plot ratio (GPR), and green cover ration (GCR) have been described in Chaps 3 and 4. The microscale greenery variables TVF GPR and GCR are plotted against measured Ta on the LCZ level, and cooling rates of inter-LCZ greenery are compared based on derived linear correlation coefficients (slope, intercept, and explanatory power, R^2).

7.1.2 Results

The daytime air temperature (3-day mean T_a, 1–5 pm) is plotted against TVF, GPR, and GCR. Simple linear-fit estimation is applied to each LCZ group. Data measured in the same period (i.e., LCZ1 and LCZ2 in the peak summer day period, and LCZ3 and LCZ4 in the later summer day period) is compared to identify impact from local effect of urbanization.

The daytime data indicates that cooling performance can be well quantified by TVF and GPR. The Ta variation at the LCZ scale can be explained by TVF, GPR in the range of 15–80% (Fig. 7.1a) and 18–80% (Fig. 7.1b), respectively. The mean R-square values of TVF and GPR are 0.57 and 0.52. As for GCR, the observed cooling rate quantified by GCR is very small and inconsistent among the data groups: The values of slope are below 0.01 in LCZ2, 3 and 4, and it even shows positive sign in LCZ1 (Fig. 7.1c).

Ambient temperature and local density/land cover have notable impacts on the cooling rate of residential greenery, as indicated by correlations of both TVF and GPR with daytime T_a (Fig. 7.1a, b): Although both measured during the peak summer period, LCZ1 was measured under a clearly hotter condition. LCZ1 has higher building density and lower greenery ratio than LCZ2, which can also contribute to a warmer state. This can be seen from values of mean T_a values and intercept of linear equations. Here, the greenery of the high-rise residential sites in LCZ1 showed a clearly higher cooling rate than that under in LCZ2. The cooling rate (slope) in LCZ1 indicated by TVF is about twice that in LCZ2; and if indicated by GPR, three times that in LCZ2. The explanatory power (R^2) is about twofold in terms of TVF (0.57 versus 0.14) and threefold (0.80 versus 0.18) in terms of GPR.

Local land cover and green ratio have notable impacts on the cooling rate of residential greenery. As for LCZ3 and LCZ4 that were both measured in the late summer day period, the ambient air temperature difference is insignificant. But the cooling rate in LCZ4 is notably higher than that of LCZ3, in the order of about twofold (slope of 0.03 versus 0.014 for TVF and 0.39 versus 0.16 for GPR). The difference in cooling rate can be attributed to that LCZ4, located at Pudong New District, has a notably higher green ratio and lower building floor area ratio than LCZ3. The higher green cover and lower density at the local scale in LCZ4 may have contributed to a better cooling performance of residential greenery at the microscale.

The scatter plots and linear-fit estimation of nighttime T_a (3-day mean T_a, 6–9 pm) with TVF, GPR, and GCR are shown in Fig. 7.2. The mean R^2 of TVF, GPR, and GCR are 0.42, 0.41, and 0.45 (excluding LCZ1), respectively. But the overall cooling rate is much lower than the daytime situation. The means of slope of TVF, GPR, and GCR are 0.005, 0.08, and 0.002, respectively, as compared to the daytime values of 0.015, 0.2, and 0.001, respectively. The local effect as shown in the daytime analysis is not clear at nighttime. The absence of solar radiation may nullify the shading effect of vegetation canopy, while the longwave irradiative cooling rate is lowered by the canopy. Also, vegetation cools mostly through evaporation at night, whereas the overall soil moisture level is low at the urban sites compared with suburb and rural area; therefore, the cooling by vegetation at night could be suppressed.

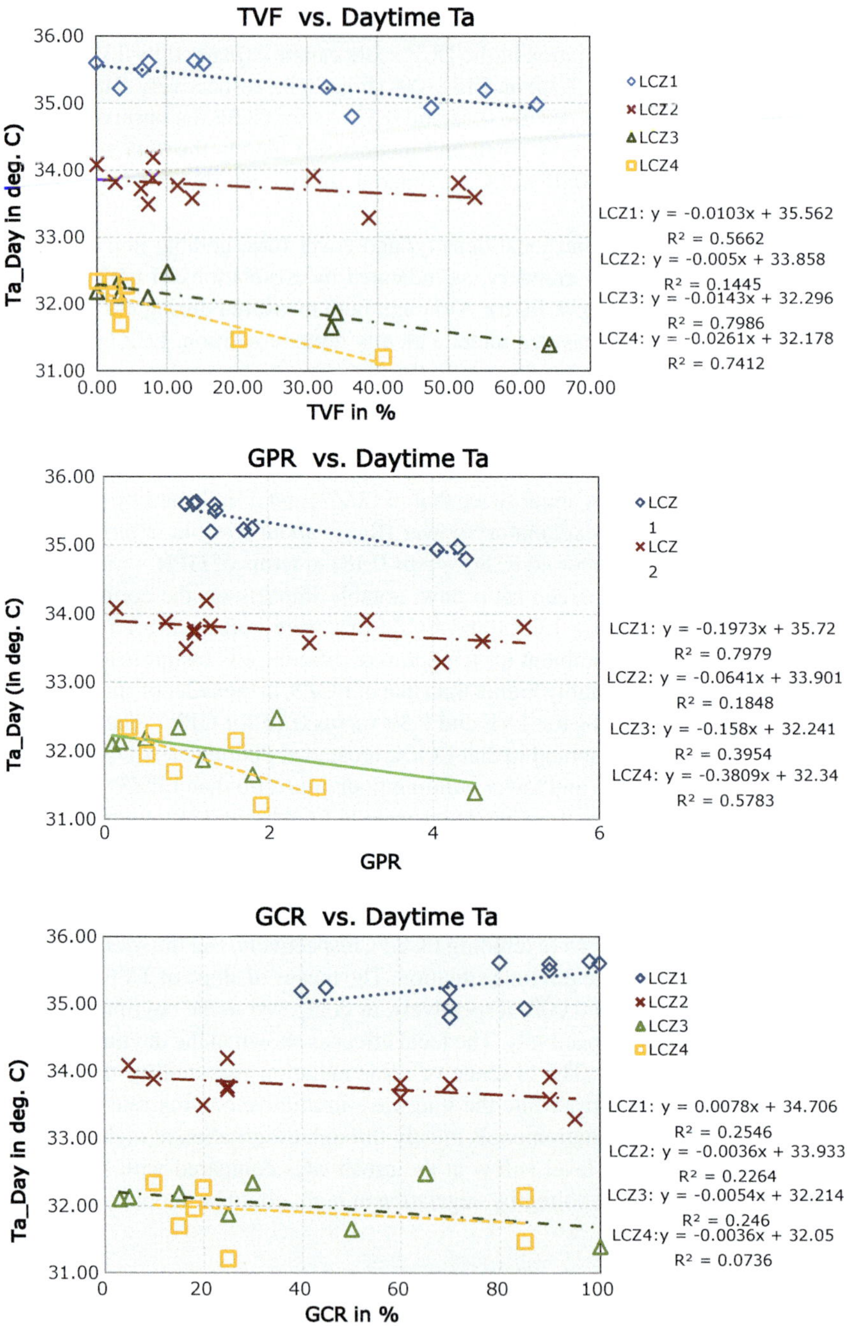

Fig. 7.1 Comparison of correlation of TVF, GPR, and GCR with T_a-day among 4 LCZs

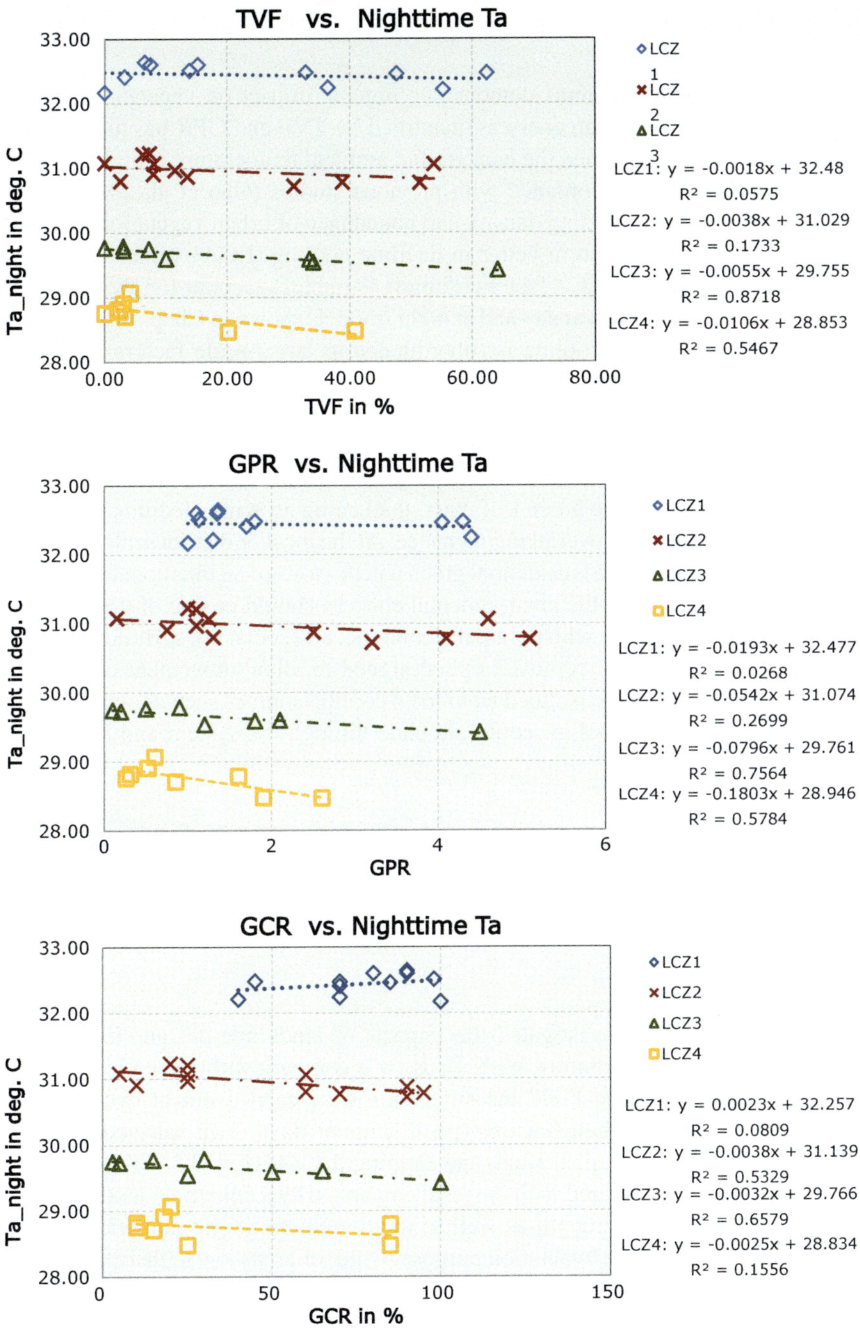

Fig. 7.2 Comparison of correlation of TVF, GPR, and GCR with T_a-night among 4 LCZs

7.1.3 Discussion

Background climatic elements demonstrate notable impact on vegetation cooling performance. Residential greenery as quantified by TVF and GPR has higher cooling rates during the day when the background ambient temperature is higher (LCZ1 vs. LCZ2). This is in accordance with previous studies (Gao et al. 2009). In an urban area with lower building density and abundance of urban vegetation, residential greenery tends to perform better in daytime cooling (LCZ3 versus LCZ4). In regression modeling, the LCZ factors (dummy variables) account for about 50% of overall UHI variation, both at day and at night (models shown in Chap. 4), indicating that residential greenery cooling is subordinated to larger-scale factors, including seasonal, geographical, and urbanization factors.

The fact that the urban land allocated to green space is very limited stresses the imperative to systemically optimize every piece of scattered residential green patches through integrated planning and design. At the microscale, landscape design should consider increasing green leaf mass, thickening growing medium, regulating irrigation, and improving overall maintenance. At the local or district scale, planning should organize the isolated residential green patches toward an interconnected green network. To achieve this, the environmental concern should be one of the priorities in decision making, along with political, economic, and social ones. Building sitting on the network should be removed or redesigned to allow movement of cool air. The network should not be isolated from local cooling sources such as sizable urban parks or lakes, so that cool air could circulate through the system and benefit the entire urban area.

7.2 The Microscale

7.2.1 Introduction

This microscale study investigated the impacts of landscape design strategies on the pedestrian thermal comfort, with focus on a comparison between high-albedo pavement and vegetation. Field measurements were carried out at two high-rise residential quarters in Shanghai on typical summer days. Two selected high-rise residential quarters are named Huali and Jingting (LCZ4) (Fig. 7.3, Table 7.1). The measured data was compared with the result generated by a numerical model, ENVI-met, to understand the strength as well as the limitations of the numerical model, before it is applied for the evaluation purpose. Outdoor areas low in thermal comfort levels which are spotted and various countermeasures are proposed, and the respective thermal comfort effects are simulated. Based on the comparison of the results, design recommendation is formulated to improve pedestrian thermal comfort.

Fig. 7.3 Photographs and site plans of the two sites with locations of the measurement points

Table 7.1 Comparison of two sites

	A (Huali)	B (Jingting)
Site layout	S-N linear	S-N linear
Site area(hectare)	8.3	4.6
Total building floor area ($\times$ 10^4 m^2)	17.7	11.8
Floor area ratio (FAR)	2.2	2.6
Green ratio (GR, in %)	53	35

The outdoor T_a distribution of the base-case scenarios of Huali and Jingting is shown in Fig. 7.4. It is clear that Huali enjoyed an overall lower T_a than Jingting. At the central spaces, the differences in paving materials and vegetation density have caused a T_a difference of about 1 °C between two sites. At Huali, P_1 and P_4 were covered with the same pavement material (ID: gs), but P_1 was about 0.3 °C lower than P_4, reflecting the distinct thermal effects by the vegetation and by the hard pavement at their windward side, respectively. In Jingting, the greenery-covered areas were lower on average by 0.5 °C than the hard-paved areas. The highest T_a occurred at the entrance plaza, which is paved with dark granite (ID: gg) and subject to the heating

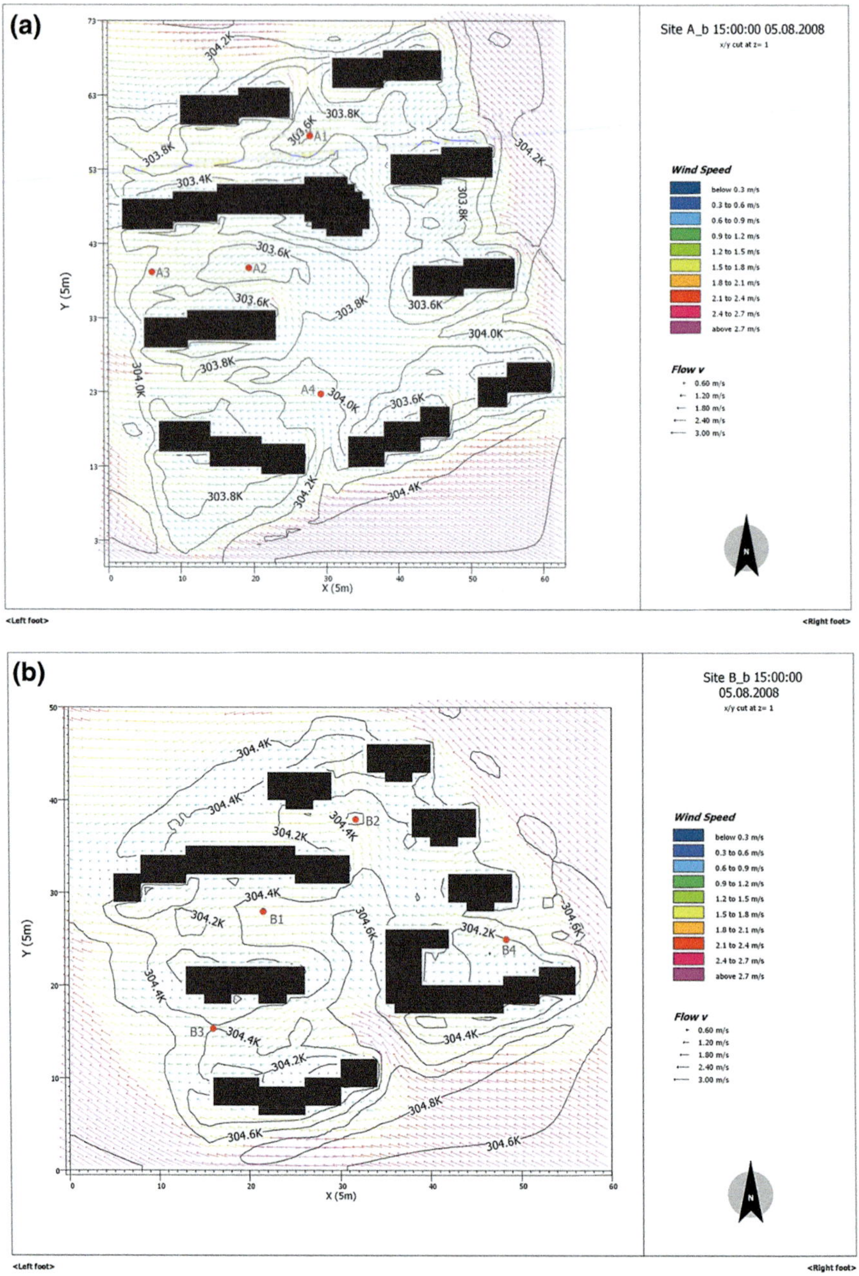

Fig. 7.4 T_a distribution at 3:00 pm at Huali and Jingting (base cases) (by ENVI-met)

effect of the asphalt roads at the windward side. Due to the adverse thermal condition in Jingting, alternative landscape design scenarios are evaluated with an emphasis on improving the case of Jingting.

The status quo scenario (base case) of two sites is simulated (see Fig. 7.4 for the 3:00 pm T_a distribution) and compared with field measurement data. The air temperature (T_a) comparison indicates that first, the effect of tree shading on air temperature reduction can be underestimated by the ENVI-met modeling. In this case about 0.5–1 °C, compared to an unshaded grass lawn; second, nighttime air temperature can be overestimated; the exact values range within 0.5–1.5 °C, depending on site conditions. For the former, the result by ENVI-met seems abnormal because shaded outdoor spaces being cooler than nearby open grass lawns on summer days are supported by a number of studies, e.g., the field study by the authors (Yang et al. 2010) and by (Shashua-Bar and Hoffman 2000). For the latter, previous research also reveals that ENVI-met tends to overestimate nighttime temperatures, and it is attributed to the static boundary conditions of the model used (Ali-Toudert 2005). The comparison has two implications: First, the evaluation based on the ENVI-met simulation has to be focused on the daytime hours; and second, the temperature reduction by a dense tree canopy (LAI = 6.4) calculated by ENVI-met in this study should be added with a compensation factor of 0.5–1 °C. The simulated wind velocities showed some disconformities at both sites. The comparison indicates that the air turbulence in high-rise built environment may not be satisfactorily modeled by ENVI-met. Nevertheless, under such a weak wind condition, WV variation is not the main factor influencing thermal comfort, in that T_a and T_{mrt} are much more important in determining the thermal balance of the human body (Hoppe 1999).

The simulated mean radiant temperature (T_{mrt}) distributions reveal reasonably the impact from urban geometry and materials. It should be noted that the simulated values of daytime T_{mrt} could be overestimated by ENVI-met because ENVI-met (Version 3.1 used in this study) considers only the attenuation effect of water vapor and neglects other particles in the atmosphere, and therefore, the solar radiation received at ground is overestimated (Ali-Toudert 2005). The amplitude of the daily T_{mrt} variation could be larger than the real situation. This is also in part due to the overestimated daytime solar radiation. In addition, the heat storage of building mass is not considered in ENVI-met (Bruse and Fleer 1998), which can lead to overestimated surface temperatures of building walls/roofs during the day and thus higher longwave radiation from the walls/roofs. Due to the same reason, surface temperatures and longwave heat emission by night can be underestimated. Another observation is that the T_{mrt} over a grass lawn seems to be underestimated by ENVI-met. Because of the limitations of the tool, it is more meaningful to pay attention to comparative results in changing design parameters with respect to the base-case scenarios instead of absolute values.

In all, five alternative urban design scenarios with differed presettings on surface albedo and greenery were simulated and evaluated (Table 7.2). The model domains of the three alternative scenarios for Jingting are shown in Fig. 7.5.

Table 7.2 Alternative design scenarios to modify thermal comfort by albedo and vegetation

Scenarios		Greenery	Pavement albedo
Jingting	J-b	The base-case scenario	
	J-1	Replace hard pavement with dense grass (ID: dg. LAI = 0.8)	–
	J-2	Cover hard pavement with large trees (ID: LT)	–
	J-3	–	Increase the albedo of hard pavement from 0.2 (ID: gs and p) to 0.6 (ID: g2)

7.2.2 *Results*

Table 7.3 lists the mean values of daytime (1–5 pm) and nighttime (6–9 pm) of air temperature (T_a), mean radiant temperature (T_{mrt}), and physiological equivalent temperature (PET) at B_1 through B_4 of the base-case scenario, as well as thermal effects (ΔT_a, ΔT_{mrt}, and ΔPET) brought about by three design improvement scenarios. It can be seen that both countermeasures of planting grass and trees (J-1 and J-2) effectively lower temperatures and improve human thermal comfort. Among the sample points, the PET reduction by replacing hard pavement with grass (J-1) at B_1 through B_3 does not exceed 10 °C in average during the daytime; but by covering pavement by trees (J-2) at B_1 leads to PET reduction by more than 15 °C. The effect of increasing surface albedo (J-3), on the contrary, significantly increases the daytime T_{mrt} and thus PET despite negligible air temperature mitigation. The PET is elevated by 2–6 °C at various points. Overall, the effect of tree planting is the best in improving pedestrian thermal comfort, secondly by grass planting and the albedo manipulation measure seems to be the worst.

7.2.3 *Discussion*

Increasing ground albedo levels worsens the daytime thermal comfort, while the improvement in nocturnal thermal comfort is marginal. An increase of 0.4 in surface albedo lowers the air temperature at the pedestrian level by 0.2–0.4 °C during the day and less than 0.1 °C at night. However, this albedo modification increases the T_{mrt} by 8–14 °C during the day, which overwhelms the relatively thinner T_{mrt} reduction by less than 4 °C at night. As a result, the overall thermal comfort suffers, as indicated by an increase of 5–7 °C in PET during the day and a decrease of less than 1 °C in PET at night.

Increasing grass cover and tree cover improved thermal comfort during the whole simulated period, i.e., from 1:00 to 9:00 pm. The modeled T_a reductions were limited, at 1 °C, but the effect of trees should be 0.5–1 °C larger than the modeled result. Nevertheless, the T_{mrt} reductions are much more significant, i.e., 12–24 °C for the

Fig. 7.5 Modeling of the two sites in ENVI-met, (Red dotted lines indicate thermally vulnerable areas that are to be improved)

Table 7.3 Thermal effects of design improvement scenarios for site B (Jingting)

Design strategies		B_1		B_2		B_3		B_4	
		Day	Night	Day	Night	Day	Night	Day	Night
J-b	T_a	33.1	30.5	32.9	30.5	32.9	30.1	32.5	29.8
	T_{mrt}	66.2	20.9	70.8	21.5	51.7	20.8	54.4	18.2
	PET	**44.3**	**23.2**	**45.9**	**23.3**	**38.2**	**23.1**	**41.3**	**22.5**
J-1	ΔT_a	−0.9	−0.9	−0.7	−0.7	−0.7	−0.6	−0.1	−0.1
	ΔT_{mrt}	−17.2	−2.9	−16.5	−3.5	−14.9	−2.7	−0.1	~0
	ΔPET	**−8.7**	**−1.8**	**−7.8**	**−1.6**	**−6.9**	**−1.0**	**−3.1**	**−0.8**
J-2	ΔT_a	−0.7	−0.8	−0.7	−0.8	−0.6	−0.6	−0.1	−0.1
	ΔT_{mrt}	−36.5	−2.8	−16.5	−3.5	−19.4	−2.5	−24.9	−0.1
	ΔPET	**−15.5**	**−0.9**	**−7.4**	**−1.6**	**−8.5**	**−1.1**	**−12.7**	**−0.4**
J-3	ΔT_a	−0.3	−0.1	−0.2	−0.1	−0.3	−0.1	0	0
	ΔT_{mrt}	7.2	−1.7	12.0	−2.2	4.3	−1.2	0	~0
	ΔPET	**3.6**	**−0.7**	**6.0**	**−0.8**	**1.9**	**−0.5**	**−0.2**	**−0.5**

Note the unit of T_a, T_{mrt} and PET are all in °C

grass scenario and 11–47 °C for the tree scenario. Consequently, PET is reduced by 2–12 °C for the grass scenario and 5–20 ° C for the tree scenario. Overall, the tree planting strategy appears to be the most effective in terms of improving outdoor thermal comfort, especially during the daytime.

7.3 The Building Scale

7.3.1 Introduction

The third case studied a double-skin green façade (DSGF) on a recently renovated office building in a university campus in Shanghai. During summer months, the thermal and biometeorological parameters of DSGF-covered rooms were measured against reference rooms, and the thermal comfort and energy-saving potentials of DSGF were estimated. The impact of weather and design parameters of the DSGF on its thermal performance is analyzed (Fig. 7.6). Two pairs of office rooms on the fourth floor (~4.2 m × 3.6 m × 3.6 m in L-W-H) were monitored. Each pair features one DSGF and one normal masonry facade (reference) with the same facade orientation. The building was constructed in the 1980s. The wall is 300 mm in width, constructed of brick and mortar, with 20 mm cement plaster, 240 mm brick, and 40 mm granitic plaster, from interior to exterior. The DSGF is planted with an evergreen plant called Mucuna sempervirens "Hemsl." One pair faces southwest south and other faces northeast north. The leaf area index is 3.0–3.5 for both the south and north green façade. Each pair was continuously monitored for around fifteen days. Measured thermal parameters include surface temperatures (T_s) of outer vegetated skin of the DSGF and masonry facade, cavity air temperature (T_a) and relative humidity (RH) in between VGS and masonry facade, and indoor thermal comfort conditions, including T_a, RH, wind velocity (WV), and globe temperature (T_g).

7.3.2 Results

- Time series of DSGF-induced bioclimatic development
  For the south pair, data collected from August 26 to September 1 is used for analysis, excluding influences from adverse climate conditions such as rains (when the thermal effect of the VGS was largely diminished). During the period of south pair measurement, the weather condition was mainly sunny and partially cloudy (Fig. 7.7a). The mean T_a is 27.2 °C, and the highest recorded T_a is 33.3 °C. The highest recorded global solar radiation (GSR) is 873 W/m^2. The mean WV is 0.4 m/s, and highest recorded WV is 3.6 m/s.
  The ambient air temperature (T_a) of the cavity between the VGS and southern building facade (with VGS) is lower than that near the bare southern facade (w/o VGS), by averagely 0.4 °C and maximally 5.5 °C during the measurement period

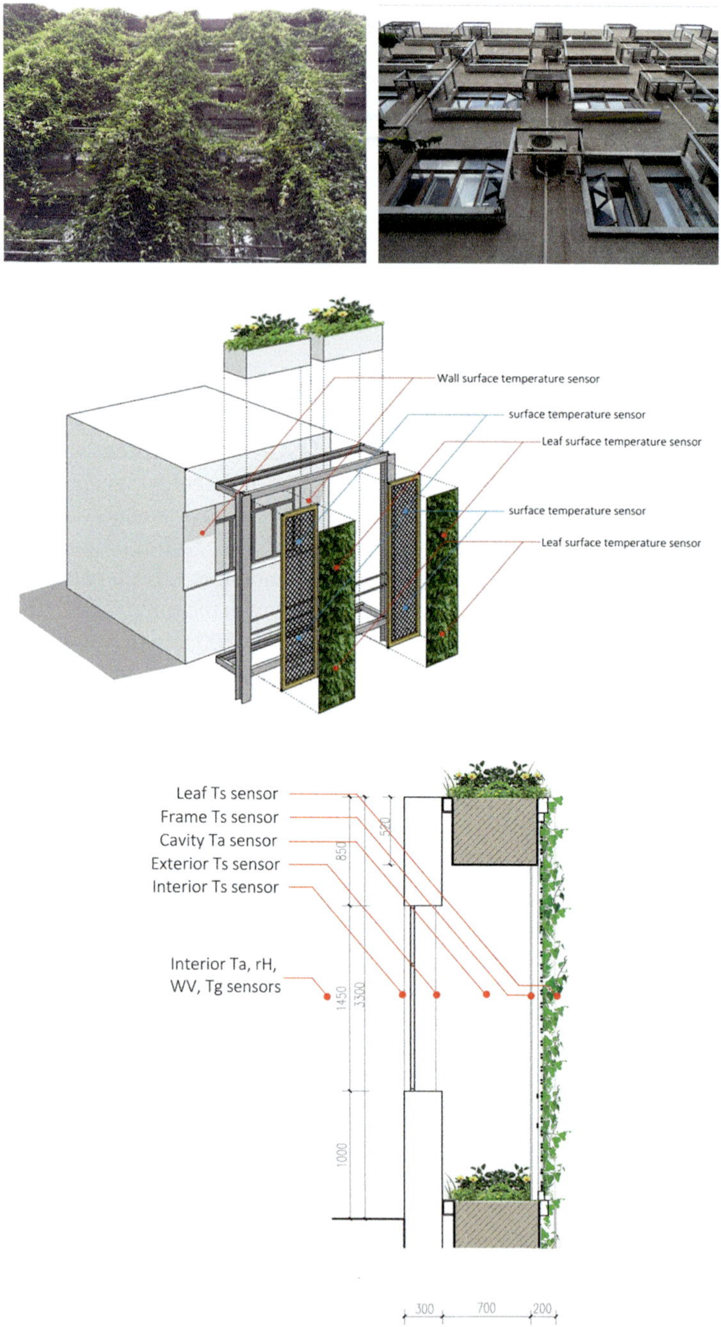

Fig. 7.6 Facade with DSGF and the original bare facade (the south pair)

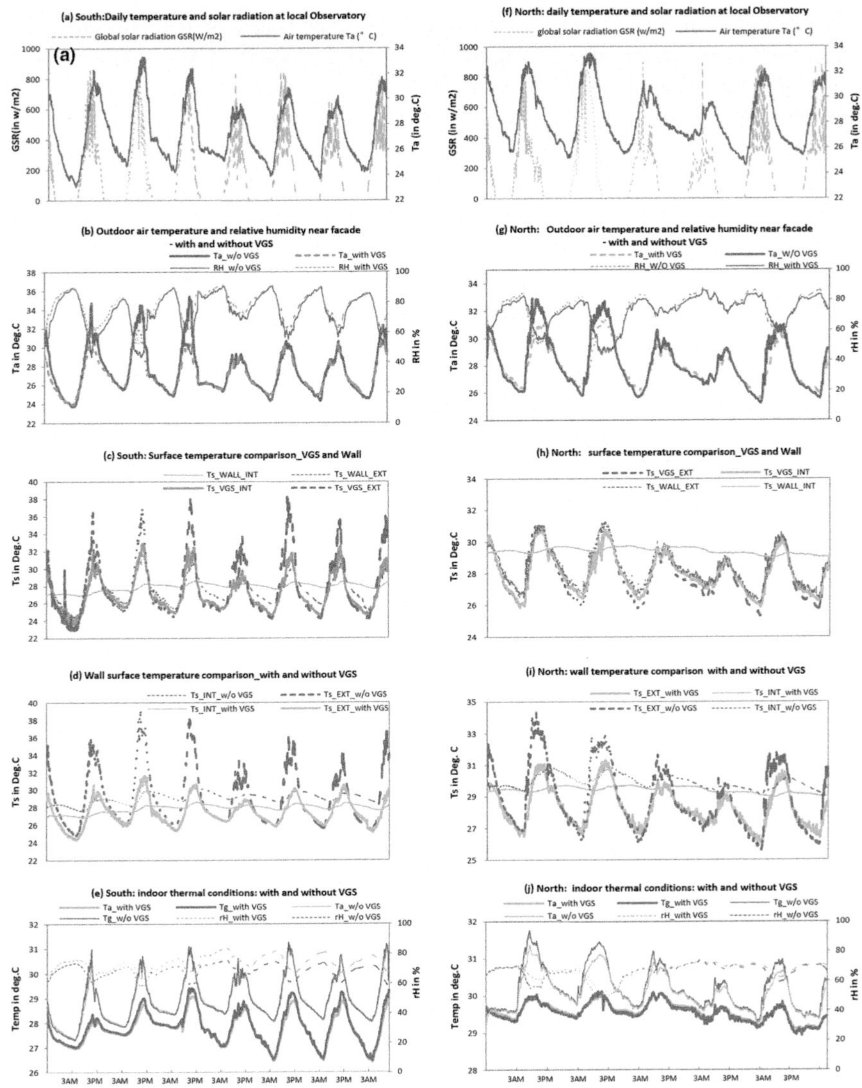

Fig. 7.7 Measurement plots on the south and north DSGF and reference

(Fig. 7.7b). The relative humidity (RH) with VGS is higher than that w/o VGS, by averagely 3.6% and maximally 22.3% during the measurement period.

The surface temperature (T_S) of the southern facade with VGS is shown in Fig. 7.7c. In average, T_S of exterior VGS is 0.4 °C higher than that of the masonry wall behind. The exterior surface of VGS is warmer than exterior surface of the masonry wall during daytime hours, by maximally 9.2 °C, but cooler in the evening and at night, by maximally −2.1 °C.

Southern facade wall T_s with and without VGS is compared in Fig. 7.7d. The mean exterior surface temperature reduction by VGS during the whole measurement period is 1.6 °C, while the maximum cooling effect reached 8.7 °C during the day, but the effect can be reversed at night under certain weather conditions (cloudy, humid, or rainy), when the VGS-covered facade can be maximally 0.9 °C warmer than that without VGS. For the interior wall surface, the facade covered with VGS is consistently cooler than that without VGS, by averagely 1.2 °C during the measurement period, and maximally 2 °C, occurred in late afternoon hours.

Indoor thermal parameters including T_a, T_g, and RH are plotted for comparison (Fig. 7.7e), while WV is not plotted because the air was generally stable, with measured WV consistently under 0.1 m/s in both rooms. In both rooms, T_g is close to T_a during most of the time, although T_g can be a little higher than T_a during the day while slightly lower at night. T_a is constantly lower in the VGS-covered room, by averagely 1.2 °C and maximally 2.5 °C; T_g largely follows the T_a pattern, with VGS-covered room averagely lowered by 1.1 °C and maximally 2.7 °C, occurred in afternoon hours. RH is consistently higher in the VGS-covered room, by 5% on average, and 9% maximum. Indoor thermal comfort is evaluated using operative temperature (T_{op}). The VGS-covered room is consistently cooler than the room without VGS. The mean T_{op} reduction is 1.1 °C, and the maximum is 2.7 °C.

The north pair analysis adopts data measured from August 13 to 17, 2015. During the period, the weather condition was mainly partially cloudy, and it was one shower of rain during August 16 (Fig. 7.7f). The mean T_a is 28.3 °C, and the highest recorded T_a is 33.5 °C. The highest recorded GSR is 896 W/m^2. The mean WV is 0.4 m/s, and highest recorded WV is 3.3 m/s.

The T_a between the VGS and northern building facade is lower than that near the bare southern facade during the day but higher at night (Fig. 7.7g). Overall, VGS made the cavity T_a averagely 0.2 °C lower than bare masonry facade, maximally 3.3 °C occurred in early afternoon. The relative humidity (RH) is on the opposite: RH with VGS is higher during the day but lower at night. Averagely VGS resulted in a little lower RH than that w/o VGS, by 0.6%, and the maximum difference is 9.2%.

The T_s pattern of northern facade with VGS is different to that of the southern facade (Fig. 7.7h). T_s of exterior VGS almost consistently remained lower than that of the exterior wall surface. During early daytime, the difference of T_s between VGS and masonry wall was very small, with the wall slightly warmer than the VGS. After 3 pm till next sunrise, the T_s of leaf surface dropped clearly lower than the VGS interior frame surface and the exterior wall. T_s difference between VGS and exterior wall surface is in average 0.5 °C during the measurement period, with maximum 1.6 °C at night.

The T_s comparison between northern-facing VGS and bare masonry facade (Fig. 7.7i) reveals that for exterior wall surface, the VGS-covered facade was clearly cooler than that of bare facade during daytime, the maximum T_s difference is 4.2 °C. The difference became smaller at night, and the VGS-covered facade became warmer than bare facade, the maximum difference can reach 1.2 °C at

night. During the whole measurement period, VGS caused a wall T_s reduction of 0.5 °C. For interior wall surface, the VGS-covered facade was consistently cooler than the bare one, averagely by 0.5 °C and maximally by 1.3 °C.
In both rooms, T_g was close to T_a most of the time (Fig. 7.7j). T_g was a little higher during the day and slightly lower at night. T_a was constantly lower in the VGS-covered room, by averagely 0.5 °C, and maximally 1.5 °C; T_g largely follows the T_a pattern, with VGS-covered room averagely lowered by 0.6 °C, and maximally 1.9 °C; RH was consistently higher in the VGS-covered room, by 1.5% on average, and 9% maximum. Regarding indoor thermal comfort, the VGS-covered room is consistently cooler than the room without VGS, the mean T_{op} reduction is 0.6 °C, and the maximum is 1.9 °C.

- Statistical comparison of VGS cooling effects
  Southern wall surface temperature reductions by VGS are plotted in Fig. 7.8b. On a daily basis, VGS generally reduced the exterior T_s of the masonry wall more than it does on the interior T_s of the wall when there were sufficiently high solar radiation and air temperature. Specifically in this case, exterior T_s reductions were only smaller than interior T_s reductions on day 4 and day 6, when GSR dropped below 300 W/m^2 and mean T_a below 27 °C (Fig. 7.8a). If consider only the working hours (i.e., 8 am–6 pm) for this office building, then the reductions in exterior T_s were significantly larger than the daily values, and they were always larger than those in interior T_s of the masonry wall, indicating the influence of solar radiation on surface temperature elevation and the shading effect of VGS, whereas the reductions in the wall interior T_s were kept at around 1.2 °C. Averagely, during the working hours, exterior T_s was reduced by about 3.4 °C and interior T_s by about 1.2 °C. While during the whole period, exterior T_s reduced by around 1.6 °C, whereas interior T_s by around 1.2 °C. As can be seen in Fig. 7.7d, the magnitude of change in the interior T_s was much lower than exterior T_s. This helps maintain a stable radiant thermal indoor environment.
  The indoor T_a and T_{op} reductions by VGS are plotted in Fig. 7.8c. The daily mean T_a reductions range 0.8–1.5 °C during the measurement period and averagely ~1.2 °C. The T_a cooling effect is a little better if consider only the working hours, nearly 1.5 °C. It is shown in Fig. 7.8d that, on a daily basis, for the VGS-covered wall, the exterior surface was cooler than the interior surface (the values being negative), the mean ΔT_s being around −0.4 °C; while the exterior surface was warmer than the interior surface for the bare wall, the mean ΔT_s being ~0.7 °C. If considering only working hours, the temperature difference was significantly higher than that of daily values for bare wall room, with the mean ΔT_s being about 3 °C, whereas for the VGS-covered room the ΔT_s is still below zero.
  The north-facing wall T_s reductions by VGS are plotted in Fig. 7.8f. Like the south-facing scenario, surface cooling effect of VGS seems positively related to solar radiation and ambient air temperature. But unlike the south-facing scenario, the cooling effects do not seem significant difference between the daily circle period and working hours only. This may be due to the very limited amount of diffuse radiation received by the northern facade. On average, T_s was reduced by ~0.5 °C for exterior and interior wall surface, for working hours and for daily circle.

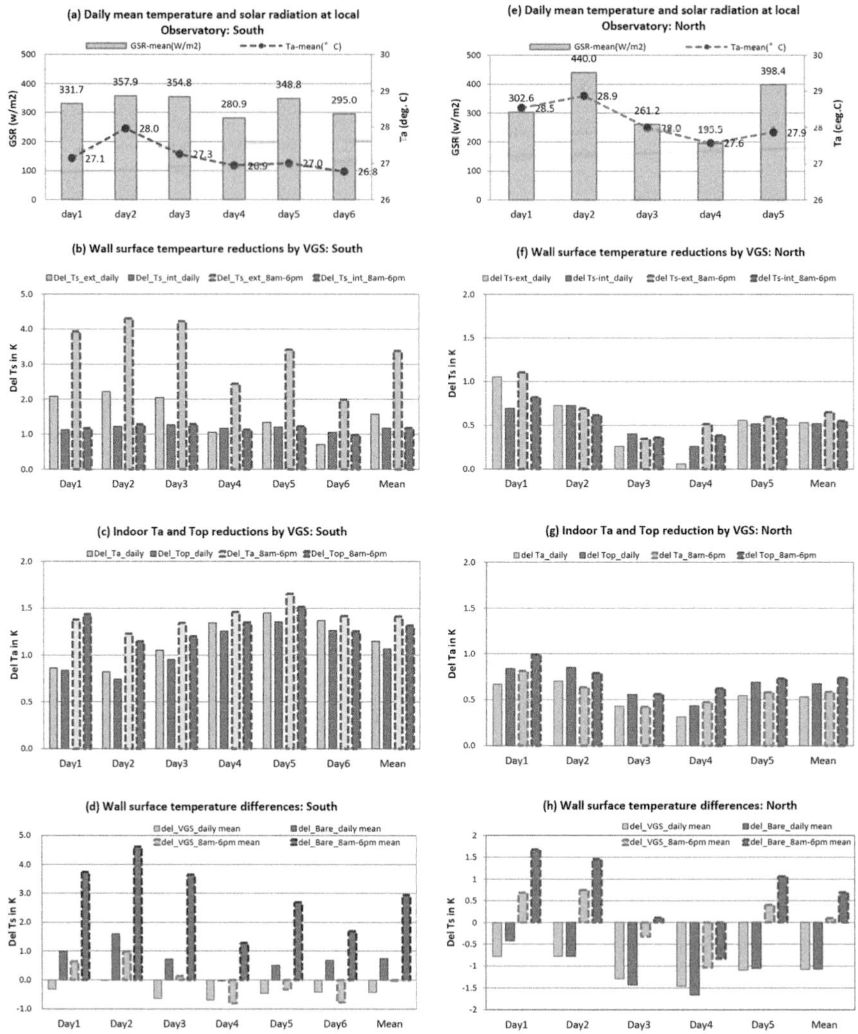

Fig. 7.8 Daily mean thermal performance of south and north DSGF

The indoor T_a and T_{op} reductions (Fig. 7.8g) were also positively related to solar radiation and ambient temperature, and the cooling range was much narrower than south-facing rooms, with about 0.5–0.7 °C for both working hours and for daily circle.

Regarding the exterior–interior ΔT_s of the north-facing wall (Fig. 7.8h), it is shown that on a daily basis, both VGS-covered and bare wall lost heat to the outside through conductive heat transfer, the mean ΔT_s being ~−1 °C. Considering the working hours, heat flux was reversed, indicating conductive heat gain, with the mean ΔT_s being ~0.1 °C for VGS-covered wall and 0.8 °C for bare wall.

7.3.3 Discussion

The VGS on the south-oriented facade created a distinctive microclimate in the cavity between the VGS and the masonry wall. The north-oriented VGS also lowered the cavity T_a but to a lesser extent. This relatively cooler air skin indicates an energy-saving benefit and a more accurate boundary for building energy simulation. When individual AC unit is used and the outdoor AHU unit is installed within this cavity, as applied in the building under investigation, the cooler air helps increase the cooling efficiency and reduce electricity consumption. A study in the Mediterranean climate suggests that the cavity air of a living wall system could be cooled by 5–6 °C, and the makeup air could lead to a cooling energy saving by 26% (Perini et al. 2017). The effect of a DSGF is unclear and can be further investigated.

T_s comparison between the VGS-covered and bare facade in the south orientation indicates a mean reduction of exterior T_s by 1.5 °C and the peak value of ~9 °C. At night, the VGS-covered wall could turn slightly warmer than the bare one, dependent on weather conditions. For the interior T_s, a mean reduction of 1.2 °C and maximally 2 °C were measured. The north pair followed the pattern, but the magnitudes of reduction were shrunk to about one half. Daily variation of the cooling effect largely followed the variation in solar radiation intensity and ambient air temperature. Note that the cooling performance of VGS on wall surfaces depends on a number of factors including vegetation characteristics (e.g., species, coverage, LAI), wall material properties (e.g., emissivity, color, heat capacity, etc.), spacing between vegetation layer and wall, in addition to orientation and weather conditions. Further study is needed to explore the effect of design variables, preferably by parametric study using validated numerical tools.

Weather condition has clear impact on the DSGF cooling effect. Higher ambient air temperature and radiation seem to have led to enhanced cooling effect of the VGS, except that the measured reductions on indoor T_a by the south-facing VGS seem negatively related to the solar radiation and ambient air temperature. Indoor warming occurred on sunny days was found in a green facade study in Hong Kong, under a subtropical climate, and the authors attributed this to the result of solar penetration through windows (Lee and Jim 2017).

Regarding the T_s differentials of facade walls, for the south pair, the exterior–interior differential during working hours was overall positive for the bare wall (~3 °C), while it was around zero for the VGS-covered wall. Similar pattern was observed for the north pair, albeit to a lesser extent. Heat flux through opaque walls is positively related to the surface temperature difference, given the same thermal resistance of the walls. Therefore, other energy transfers being equal (i.e., radiation through window and convection through fenestrations, etc.), and assuming no change in stored heat of the masonry wall, the exterior–interior T_s difference can give a proximate estimate of cooling energy load. So, theoretically indoor T_a reduction and T_s differentials both indicate potential cooling energy savings (Eumorfopoulou and Kontoleon 2009; Perini et al. 2017).

More comprehensive measurement and simulation are needed before any conclusion on energy impact of a VGS can be drawn. Radiation and convective heat transfer through windows need to be considered. The shading effect of VGS on indoor thermal and illumination conditions should also be taken into account. The overall energy performance should be assessed taken the above-mentioned factors into account.

7.4 Conclusion

The first case study looked into the cooling performance of the residential greenery in localized urban climates. The inner city is divided into four LCZs, and the micrometeorological as well as greenery parameters were measured from a set of ten residential quarters throughout four LCZs. The bivariate models using TVF, GPR, and GCR are capable of explaining about a half of the measured variation in air temperature, within each LCZs. The results support that residential greenery cooling can be subordinated to local-scale seasonal, geographical, and urbanization factors. The analysis on the collected data made it possible to gain a first-hand understanding on the microscale vegetation cooling patterns under the local urban effects.

The second case studied the thermal comfort effects of various urban design strategies and focused on the application of high-albedo materials and greenery. ENVI-met was applied to numerically simulate and evaluate different design scenarios varying in surface albedo and vegetation parameters. It is found that increasing greenery cover, especially tree cover, improves thermal comfort during the whole period under evaluation. In contrast, increasing ground surface albedo overall worsens the thermal comfort, as indicated by significant increases in PET during the day with a marginal decrease of PET at night.

In the third case, the thermal environmental parameters and biometeorological indices were measured and the impacts on cooling energy use and indoor thermal comfort analyzed for a double-skin green facade (DSGF) system against the original bare masonry facade. The VGS on the south-oriented facade created a distinctive microclimate in the cavity, indicating a theoretical energy-saving benefit and a more accurate boundary condition for energy simulation of buildings with VGS. The air temperature averagely dropped by about 0.4 °C in a daily circle and reached peak value of 5.5 °C. The north-oriented VGS lowered the cavity T_a to a lesser extent. Surface temperatures of VGS and the wall behind it revealed a positive gradient from exterior to interior during the day, while this gradient was largely reversed at night with the absence of solar radiation. The south VGS reduced exterior wall T_s by 1.5 °C in average and ~9 °C maximally. Indoor air temperature reduction of averagely 1.2 °C was observed for the south-oriented VGS-covered room. The surface temperature differentials of facade walls during working hours indicate reduced cooling load for VGS-covered rooms.

References

Ali-Toudert F (2005) Dependence of outdoor thermal comfort on street design in hot and dry climate. *Meteorological Institute*. Feiburg, Freiburg University.

Bowler DE, Buyung-Ali L, Knight TM, Pullin AS (2010) Urban greening to cool towns and cities: A systematic review of the empirical evidence. Landscape Urban Plann 97:147–155

Bruse M, Fleer H (1998) Simulating surface-plant–air interactions inside urban environments with a three dimensional numerical model. Environ Model Softw 13:373–384

Che S, Song Y (2001) Extract of the remote sensing message of urban green space landscape—Shanghai city as the Case Study (in Chinese). Urban Environ Urban Ecol 14:10–12

Eumorfopoulou EA, Kontoleon KJ (2009) Experimental approach to the contribution of plant-covered walls to the thermal behaviour of building envelopes. Build Environ 44:1024–1038

Gao K, Qin J, Song K, Hu Y (2009) Fallen temperature effects at green patches of urban residential areas and analysis of its influence factors. J Plant Resour Environ 18:50–55 (in Chinese, with English summary)

Hoppe P (1999) The physiological equivalent temperature—A universal index for the biometeorological assessment of the thermal environment. Int J Biometeorol 43:71–75

Lee LSH, Jim CY (2017) Subtropical summer thermal effects of wirerope climber green walls with different air-gap depths. Build Environ 126:1–12

Oke TR, Crowther JM, McNaughton KG, Monteith JL, Gardiner B (1989) The micrometeorology of the urban forest. Philos Trans R Soc London Ser B Biol Sci 324:335–349

Perini K, Bazzocchi F, Croci L, Magliocco A, Cattaneo E (2017) The use of vertical greening systems to reduce the energy demand for air conditioning. field monitoring in Mediterranean climate. Energy Build 143:35–42

Shashua-Bar L, Hoffman M (2000) Vegetation as a climatic component in the design of an urban street: an empirical model for predicting the cooling effect of urban green areas with trees. Energy Build 31:221–235

Yang F, Lau SSY, Qian F (2010) Summertime heat island intensities in three high-rise housing quarters in inner-city Shanghai China: Building layout, density and greenery. Build Environ 45:115–134

Yang F, Yuan F, Qian F, Zhuang Z, Yao J (2018) Thermal and energy performance of a double-skin green facade: A case study in Shanghai. sustainable cities and society 39:43–51. https://doi.org/10.1016/j.scs.2018.01.049

Yang F, Lau S, Qian F (2015) Cooling performance of residential greenery in localised urban climates: A case study in Shanghai China. Inter J Environ Technol Manag 18(5/6):478–503

Chapter 8
Microclimate in High-Rise Central Business Districts

Abstract *Study intent* Central Business district features intensive land use, diverse building form and functions, high volume of vehicle traffic, lower greenery ratio, and of course taller buildings, compared to high-rise residential districts. Thermally comfortable outdoor environment would bring economical and social benefits to the area; so, it is important to understand and evaluate the impact of development on its microclimate and pedestrian comfort. We developed a thermal atlas methodology for this purpose. The thermal atlas is based on empirical modeling, digital elevation model data processing, and spatial analysis. Key morphological indicators were used to account for six aspects of urban climatic impact, i.e., building density, land parcel use, anthropogenic heat, greenery, ventilation potential, and heat sink. The resultant thermal atlas comprises the following components: urban morphological maps (or sub-layers); empirical models (for sub-layer weighting and model validation); thermal comfort indices maps; and thermal zoning and design recommendation maps. Shanghai Lujiazui CBD, the culmination of CBD development in China, is assessed as a case study. The second case studied the Lujiazui Elevated Walkway (LEW) to complement the application of thermal atlas system at the ground level. In high-rise commercial districts, elevated walkway system is becoming an indispensable way to connect plots and buildings separated by driveways and avenues. Assuming more breezes on elevated levels compared to sidewalks at the ground levels, walking high can be, however, exposed to higher solar radiation and thus higher radiant temperatures without proper shading. The case study aims to gain an empirical understanding of the overall effect of changing in elevation on pedestrian summertime comfort. *Results and discussion* Based on the results, design suggestions are made for Lujiazui CBD, i.e., providing opaque shading devices for major pedestrian spaces at century walkways and waterfront esplanade; reducing the size of street blocks; dividing massive single buildings into building clusters with smaller spacing; and improving the accessibility to the heat sinks. The thermal atlas can rapidly analyze and visualize urban microclimate variations as affected by different urban design scenarios, thus a useful decision-support tool. For the LEW case study, data analysis based on the biometeorological measurements and guided questionnaire survey indicates that, the LEW was more uncomfortable than the ground level during the measured period: air

The content of this chapter is published in: Yang and Chen (2016), Yang et al. (2016).

F. Yang and L. Chen, *High-Rise Urban Form and Microclimate*,
The Urban Book Series, https://doi.org/10.1007/978-981-15-1714-3_8

temperature was higher, but wind velocity is lower on the skywalk level than on the ground level, which is counterintuitive. It could be due to the convection enhanced by buoyancy between shaded and unshaded places. The resultant thermal comfort index indicates warm conditions on the ground level (when shaded) whereas hot conditions on the skywalk level. Countermeasures of various shading design and evaporative (mist) cooling design are discussed so as to improve thermal comfort level.

8.1 Thermal Atlas of Lujiazui CBD

8.1.1 Introduction

Urban designers normally make environmental judgment based on their perception of the region, city and site, as well as their knowledge and experiences. Nowadays, as cities keep growing larger and denser, designers are constantly faced the challenges of working in high-density and high-heterogeneity urban environment. In such, tools are in great need to help designers gain a better microclimatic understanding by providing predictive results for different design scenarios with reasonable resolution and accuracy. The commonly employed tools are based on computational fluid dynamics (CFD) (Chen 2004). CFD tools require expertise to operate, are generally resource-costly, and thus unfamiliar and not user-friendly to architects and urban designers. On the other hand, at the urban scale, urban climatic mapping (UC-Map) emerges as a promising decision-support tool that aims to provide urban planners with key analytical climate information in a very "visual" form (Chen et al. 2012). UC-map studies have been carried out in a number of cities in Europe, North America, and Asia. Among them, the project commissioned by the German government may be the first effort in integrating and legitimating climate information into urban planning process. On the metropolitan and urban scale, the urban form-based variables are generally categorized into two layers of Thermal Load and Dynamic Potential; regional wind information was also provided (VDI. 1997). The established paradigm has been adopted in various cities in Germany and has influenced subsequent UC-Map research and application across the world (Tokyo Metropolitan Government TMG 2005; Planning Department Hong Kong SAR 2009).

Earlier UC-Maps generally stratify climate information based on land use and rely largely on expert evaluation. A more recent UC-Map in Hong Kong factors building density, topography, ground roughness and greenery into street-level urban microclimate evaluation (Ng 2012), and a systematic and quantitative approach is taken, using urban morphological variables such as sky view factor (SVF) and frontal area density (FAD) to describe thermal and aerodynamic characteristics of high-rise high-density urban environment of Hong Kong (Chen and Ng 2011). Our previous work indicates that thermal variations at pedestrian level can be well explained by empirical models built on carefully selected urban morphological variables (Chaps. 4

and 5). The thermal atlas are considered more quantitative than the previous UC-Map systems, by weighting the effects of urban design parameters into regression modeling and map generation. It can be one way to advance the UC-Map analytical system with respect to its application in the microscale urban climate analysis suitable for urban design practice.

- The Lujizzui CBD

The Lujiazui CBD is located in the Pudong New District on the eastern bank of Huangpu River just across the Bund. The total land area is 1.7 km^2 and covers commerce, business, and high-end residential land uses. The total building floor area is about 4.35×10^6 m^2, and the gross floor area ratio about 2.5. With totally more than 40 super high-rise buildings (building height >200 m), Lujiazui (LJZ) CBD is considered as one of the most important financial hubs in China, and its planning and design have strong influences on the development practice of other Chinese cities. It is therefore selected as the case study area. Figure 8.1 shows the map of LJZ CBD and its digital elevation model (DEM).

The study follows the framework of Thermal Load and Dynamic Potential in classic UC-Map system, but classifies the urban microclimatic characteristics into six categories, i.e., building form/density, land use/cover, anthropogenic heat, vegetation cooling, ventilation potential, and proximity to heat sink. Based on the previous work of authors and a systematic review on the state of the art in urban climatology and urban morphometry, the key morphological parameter is identified for each category, i.e., sky view factor (SVF), frontal area density (FAD), green plot ratio (GPR), hard pavement ratio (HPR), proximity to heat sink (PHS), and vehicle travel density (VTD) (refer to Chap. 3 for details of the parameters).

- Thermal comfort assessment

The biometeorological index of physiological equivalent temperature (PET) is used to assess the overall effect of the local microclimatic condition on human thermal comfort. PET is a temperature scale (in °C) index that enables the assessment of the thermal components of climates based on personal experience. It is defined as "the air temperature at which, in a typical indoor setting, the heat balance of a human body is maintained with core and skin temperatures equal to those under the conditions to assess." (Hoppe 1999). By definition, PET builds a linkage between complex outdoor climatic environment with the standardized indoor thermal environment and experience. The unit is degree Celsius, same as air temperature, the effect of which can be readily evaluated based on daily experience and common sense. In addition, the temperature scale makes it easy to be interpreted by individuals without advanced knowledge and expertise in physiology or meteorology, which makes it commonly accepted among designers and planners. Originated in mid-high latitude Germany, PET has been applied in a number of thermal comfort studies in different climates including hot–humid (sub-) tropical regions, e.g., Taiwan (Lin et al. 2010) and Hong Kong (Niu et al. 2015).

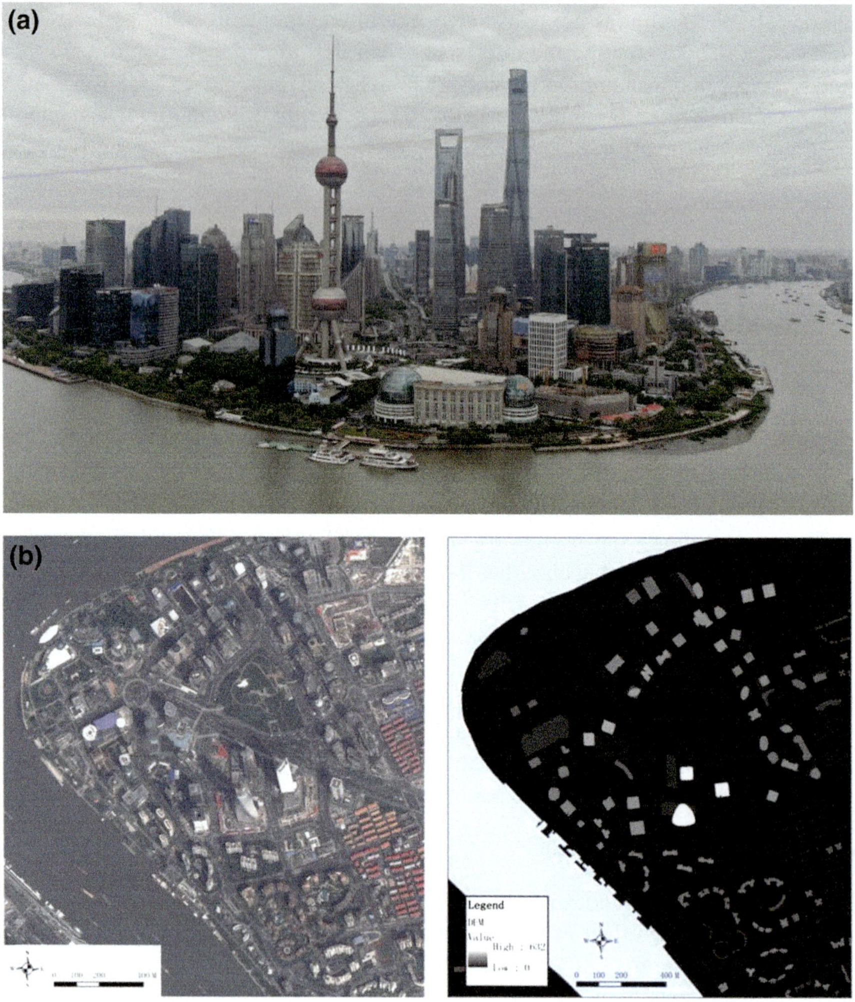

Fig. 8.1 View of Lujiazui CBD from northeast (**a**) (Photograph credit: Minggang Yin, used with permission) and a satellite map (Baidu Maps) and a DEM (**b**)

- Biometeorological measurement

Summertime field measurement campaigns were carried out in July 2013. A fixed weather station was installed at 1.5 m above ground level in the open lawn of the Lujiazui Central Green (LCG), continuously recording wind velocity (WV), wind direction (WD), air temperature (Ta), relative humidity (RH), and global solar radiation (GSR) at a 10-second interval. A multi-day synchronized traverse measurement was carried out along six preselected pedestrian routes. Each route links nine measurement points, started from the LCG and ended at the waterfront area. A tailor-made portable weather tracker was carried by two research assistants, measuring WV, WD,

Ta, RH, and globe temperature (Tg) at the 10-second interval. The six measurement routes cover 54 measurement points. Mean radiant temperature (T_{mrt}) was calculated based on Ta, RH, WV, and Tg according to the methods suggested by (ISO-7726 1998). The selected thermal comfort index PET was calculated using Ta, RH, WV and T_{mrt}, and assuming constant metabolism rate of 80 W and cloth levels of 0.9 clo (clothing level) (Hoppe 1999).

8.1.2 Results

- Biometeorological measurement

The traverse and fixed measurement points are shown in Fig. 8.2. The weather conditions in the two days are fairly similar, partially cloudy, and hot and humid.

Fig. 8.2 Map of Lujiazui CBD showing the traverse and fixed stations

The prevailing winds show a clear southeastern pattern, and an overall weak wind environment at pedestrian level.

Site photographs, hemispherical sky images, and meteorological data profiles of measurement points along Route A are shown in Fig. 8.3. It can be seen that the temperature variation, particularly the globe temperature (Tg), was largely affected by the degree of openness to the sky. For instance, the tree-shaded point A5 was remarkably cooler than adjacent locations. The air temperature (Ta) was even lower than the reference LCG station during both morning and afternoon measurement sessions because of the dense tree shading, Tg has decreased approaching Ta, indicating improved thermal comfort compared to points under higher degrees of solar exposure.

- Empirical modeling

The measured SVF showed positive correlation with PET: it can explain more than 50% of the variations in PET (Fig. 8.4), which confirms that SVF is an important influencer of thermal comfort. On the other hand, ground material did not show a decisive impact on the local thermal condition: for instance, a point over grass lawn (A8) was hotter than the next point over a paved waterfront esplanade (A9).

The empirical datasets are divided into two groups: modeling group and validation group. Each group has 27 measurement points. The empirical modeling process was carried out based on the modeling group datasets. Pearson correlation and linear-fit regression were carried out among the thermal indices PET, six morphological variables, and three control variables, namely Ta_cg, Vw_cg and SR_cg which are air temperature, wind velocity and solar radiation at the reference LCG station, respectively. The result indicates that, among the six morphological variables, PET is significantly related with SVF, GPR, FAD, and PHS at 0.01 significance level (2-tailed), and related with HPR at 0.05 level (2-tailed). For the control variables, PET is only significantly related to Ta_cg at 0.05 level (2-tailed). A nonparametric correlation test (Spearman's rho) reveals similar results.

The analysis also reveals that VTD is insignificant in the correlation. We speculate that this is probably because that VTD is assumed to map the spatial distribution of traffic-induced waste heat during a relatively longer period, whereas the measurements only captured the heat from nearby vehicles during short periods (i.e., 10 min). In this sense, long-term measurements with fixed weather stations are likely to better measure the impact of VTD, and on the other hand, vehicle-counting could be an alternative for a short-term measurement, as suggested in (Hii et al. 2015). This has to be considered in future work to improve the experiment design.

When building a multiple linear regression model, one must check the potential collinearity problem among independent variables. In this case, SVF is found to be significantly correlated with FAD, HPR and GPR. The negative correlations of SVF with FAD and GPR are reasonable, as increasing area of building facade generally reduces the viewable sky accordingly, and increasing coverage of vegetation, especially trees reduces it as well. The negative correlation occurred between SVF and HPR could be attributed to that the measurement points with lower SVF are generally located between tall commercial buildings where the ground surface tends to

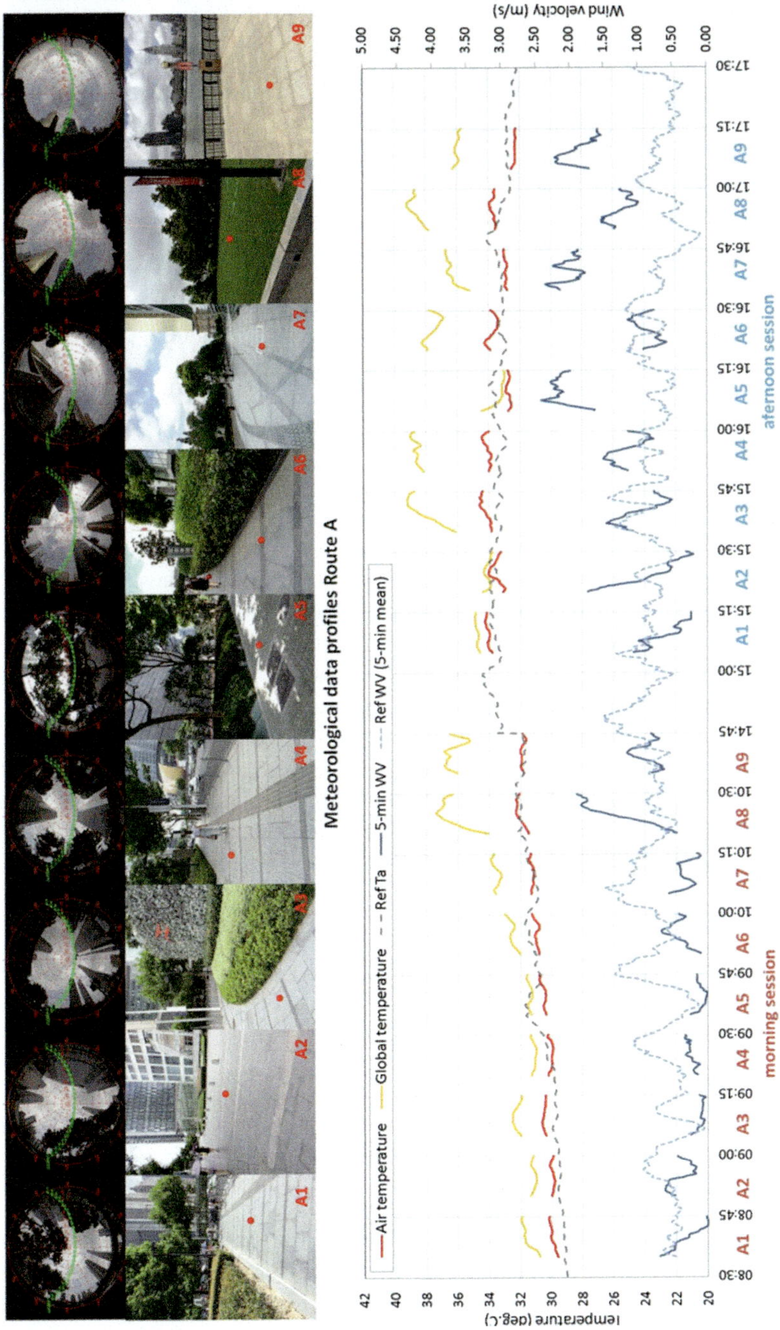

Fig. 8.3 SVI and photographs of measured points on Route A, and measured data profiles

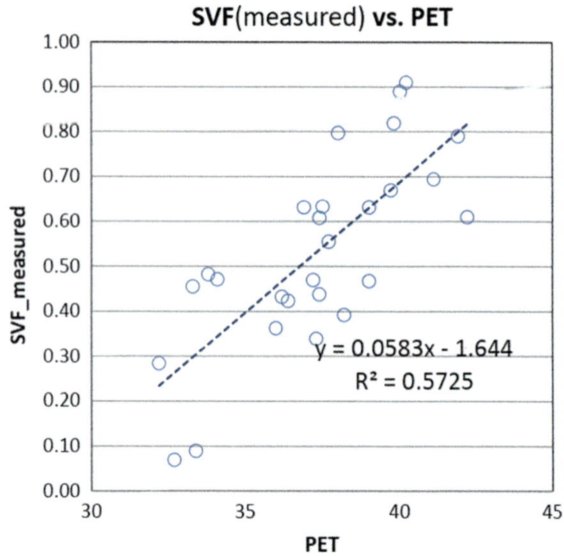

Fig. 8.4 Correlation of measured SVF with PET

be impervious (roads, sidewalks and parking lots, etc.), whereas those with higher SVF are mostly located within or close to open spaces (small parks or waterfront greenery) and consequently less hard surfaces are found around them.

With the collinearity effect in mind, regression analysis was carried out including all the six morphological variables and one control variable (i.e., Ta_cg). Significant variables were firstly all entered to obtain a model with all sub-layers of variables and with the highest explanation power. Less important variables were then removed (Stepwise, Backward and Forward Methods), in order to build a parsimonious model that keeps the key variables while retaining the highest possible explanatory power. The result is as follows (Eqs. 8.1, 8.2 and 8.3). (ITD: Intra-urban Ta differential):

$$\mathrm{PET} = 8.51 + 0.81 \cdot \mathrm{Ta\_cg} + 3.66 \cdot \mathrm{SVF}{-}0.31 \cdot \mathrm{HPR}{-}1.07 \cdot \mathrm{GPR}{-}4.13 \cdot \mathrm{FAD} + 0.004 \cdot \mathrm{PHS}\left(R^2 = 0.79\right) \quad (8.1)$$

$$\mathrm{PET} = -48.33 + 2.58 \cdot \mathrm{Ta\_cg} + 7.42 \cdot \mathrm{SVF}—0.93 \cdot \mathrm{GPR}\ \left(R^2 = 0.76\right) \quad (8.2)$$

$$\mathrm{ITD} = 20.39—0.64 \cdot \mathrm{Ta\_cg} + 0.85 \cdot \mathrm{SVF}—0.16 \cdot \mathrm{GPR}\ \left(R^2 = 0.62\right) \quad (8.3)$$

- Model verification and validation

The validation group is used to test the validity of the models. The PET values calculated using field-measured data are plotted against modeled PET values. The scatter plots and linear-fit estimations are shown in Fig. 8.5 the linear estimation fits fairly well, and Eq. 8.2 shows a higher explanatory power (~0.64) than that of Eq. 8.1 (~0.54). This is reasonable, because considering limited the sample size,

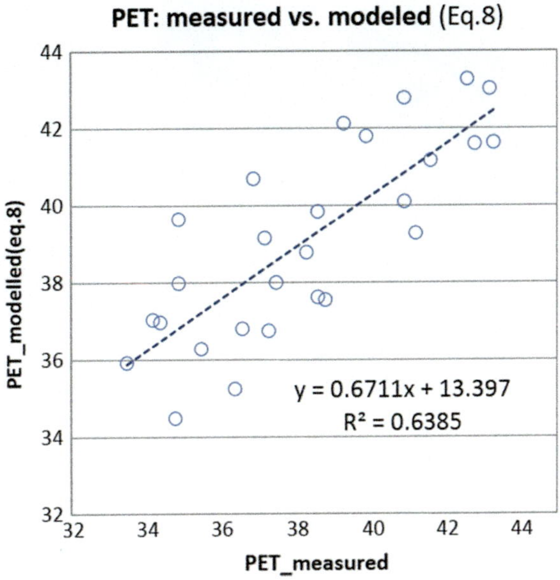

Fig. 8.5 Linear fit of field-measured PET with modeled PET

the simpler model with less independent variables is more robust (Eq. 8.2, with *F* statistics = 23.6) than the model with all important variables (Eq. 8.1, with *F* statistics = 10.3). The validity of the built model in explaining thermal comfort level with morphological variables is therefore confirmed.

- Generating the thermal atlas

The morphological sub-layers were generated, as shown in Fig. 8.6. The PET maps were than generated based on the sub-layers and the empirical modeling results. The reference air temperature (Ta_cg) is set as a constant, using daytime mean Ta of 32.1 °C during the measurement period at LCG station. Figure 8.7 shows the PET and ITD maps generated based on the two regression results (Eqs. 8.2 and 8.3).

The PET mapping is synthesized using the raster function in GIS, and the contributions from respective variables are weighted based on the regression coefficients. According to the range of PET values, six classes are assigned referring to the ranges of PET for different grades of thermal comfort perception and thermal stress given by (Matzarakis et al. 1999), Class 1 and 2 can be categorized as warm, Class 3–5 as hot, and Class 6 as very hot.

The PET map indicates that, in the summer days, the overall thermal condition in LJZ Shanghai can be ranging from warm to hot. The spatial variations of PET are mainly due to the differences of building density and layout, vegetation and proximity to heat sink.

The warm area (Class 1 and 2) is mainly distributed in densely vegetated area, i.e., LCG, Lujiazui Waterfront Park and several high-end housing quarters at the southern LJZ area. Smaller patches exist around commercial buildings, due to the low SVF caused by building shading. The street blocks to the north side of LCG primarily

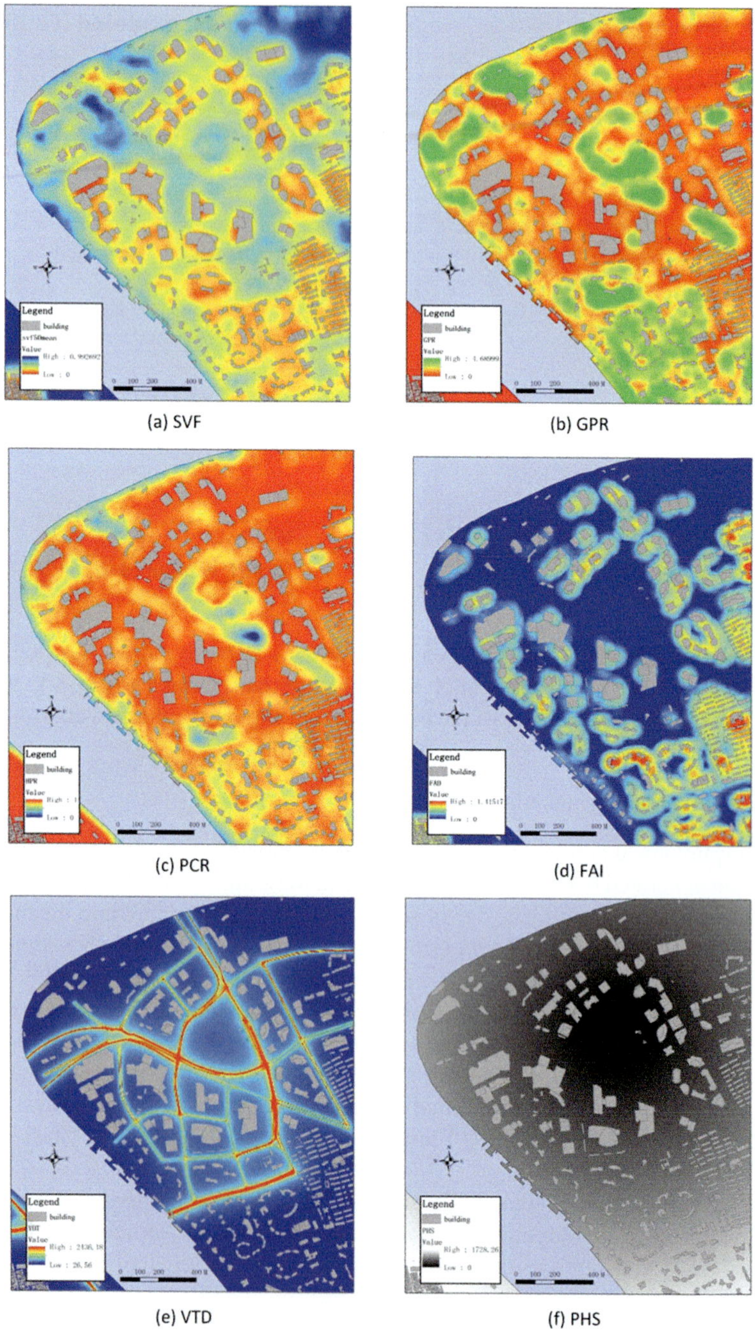

Fig. 8.6 Six morphological sub-layers (50 m radius average)

Fig. 8.7 PET and ITD map f (Ta_cg, SVF, GPR)

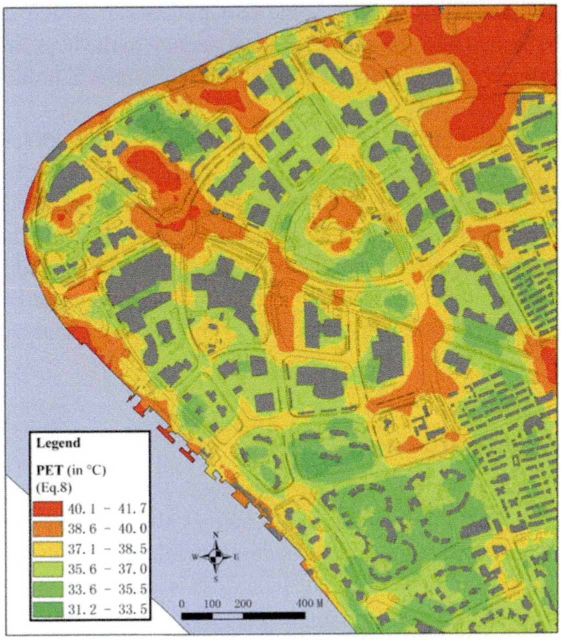

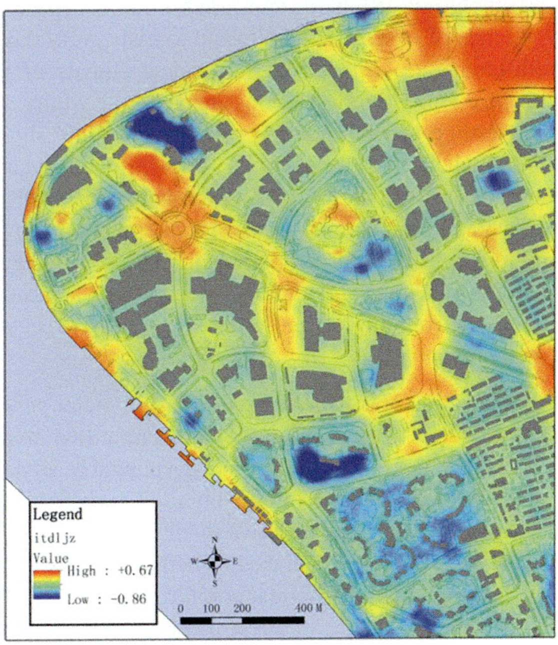

fall into Class 2, with some part in Class 1, which can be attributed to shading of buildings and street trees, and adjacency to the heat sink, i.e., LCG at their windward side. The local thermal stress level is moderate, indicating walkable outdoor spaces during the day.

On the other hand, the hot and very hot areas (Class 3–6) area normally found at wide roads, open grass lawn, and paved plazas. In particular, the area in the northeast of investigated area is found to have the most severe heat stress. In fact, it is a construction site (as can be seen from Fig. 8.2), and therefore, the very hot condition is largely caused by high levels of SVF and HPR, and low level of GPR. The updated morphological and land cover information should be added to generate new PET maps when the construction is completed to reanalyze and re-evaluate that area.

8.1.3 Discussion

Urban planning and design practices always strives to strike a balance among many issues and requirements from political, economic, social as well as environmental aspects. In this case study, LJZ CBD is one of the most prestigious financial hubs and tourist attractions in China. Therefore, the planning and design of LJZ CBD should give emphasize on making comfortable outdoor spaces which can benefit hundreds of thousands of office commuters and tourists every day. Unlike in mid-latitude cities with temperate climate, the subtropical climate of Shanghai requires that design prioritize summertime heat dissipation over winter shelter/heating. The generated summertime thermal comfort map in this study should be revealing to designers, policy makers, and all stakeholders, as it provides clear visual information on outdoor thermal comfort levels, as well as its relationships with urban morphological variables which can be moderated by design intervention. To simplify the hierarchy system and highlight the most important information, we generate a thermal zoning map based on PET map and identify three thermal zones (TZs) (Table 8.1 and Fig. 8.8)

Judged from the thermal zoning map, three aspects are worth considering design for improvement:

(1) Providing opaque shading devices for major pedestrian spaces at century walkways and waterfront esplanade. These two areas are highlighted in Thermal Zone 1, i.e., the most thermally vulnerable areas. Both places are designed for pedestrians and are intensively used by tourists and office commuters, but currently without any effective shading devices or tree cover. To add sun protection at these places will be the most effective measure that can immediately improve summer daytime thermal comfort.
(2) Reducing the size of street blocks, and dividing massive single buildings into building clusters with smaller spacing. It can be observed from Fig. 8.8, that the street blocks with smaller sizes and building clusters with smaller footprint (but closer to each other) are significantly cooler than those larger street blocks with a single and massive building. This is because denser urban fabric creates

Table 8.1 Thermal zoning map and design recommendations

Thermal zones and features	Urban morphological characteristics	Measures for improvement
TZ1: extremely vulnerable to heat stress	Open area with little obstruction to sky, lowest vegetation coverage: paved waterfront; construction site; parking lots; major road intersections	Provide shading (man-made canopy or tree cover where appropriate); increase vegetation coverage; enhance proximity to heat sink
TZ2: high density and high usage space	Large street blocks; massive buildings with large footprint; large building spacing; grass lawn with no tree coverage; major roads and pavements	Reduce the street block scale; replace large single buildings with smaller and adjacent buildings to create shadowed space and wind path; increase vegetation coverage
TZ3: climatic valuable area	Green space with dense tree cover, e.g., LCG and waterfront park; residential quarters with dense vegetation; tall buildings with small spacing and smaller street blocks	Maintain and improve

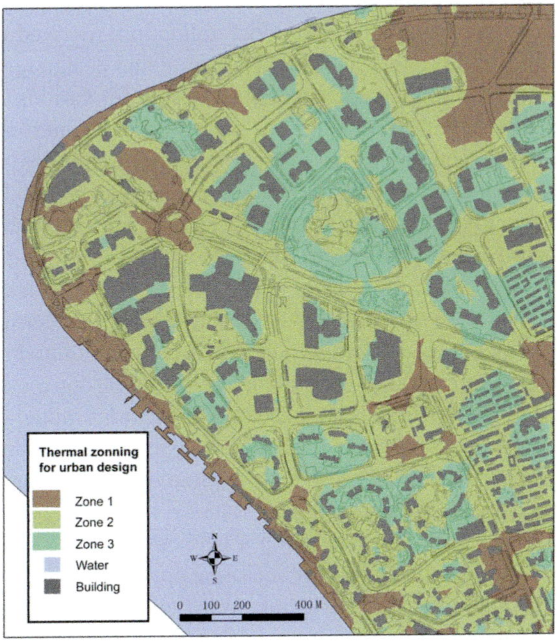

Fig. 8.8 Thermal zoning map based on PET and ITD map

shaded outdoor spaces between buildings, whereas isolated super-tall or super-large buildings generally occupy larger street blocks (typically >400 m for each side of the block) and are separated by wide roads. This leads to uncomfortable pedestrian environment, because of less solar protection, and more sensible heat from nearby road pavement materials. Refining the urban grid and building fabric to a scale of 150–250 m not only improve daytime thermal comfort, but also make the city more walkable and thus more pedestrian-friendly.

(3) Improving the accessibility to the prominent heat sinks. For instance, expanding the existing walkway system to improve the connection between surrounding office buildings with LCG, so as to increase the occupation and usage of LCG during lunch hours and weekend days; designing building and landscape elements carefully, for instance, by reorganizing short walls, trees, etc., so as to introduce more cool breeze from LCG to its surrounding built-up areas. Also, overly wide roads such as the Century Avenue in Lujiazui CBD (100 m in width) should be avoided, because on one hand, it is very pedestrian unfriendly; on the other, the street blocks on the southwest side of the Century Avenue could enjoy better cooling effect from LCG with a narrower road with lower heat capacity.

The case study tested the methodology of combining empirical modeling and spatial analysis techniques in creating a thermal atlas system at the urban district scale. This thermal atlas system comprises the following components: urban morphological maps (or sub-layers), empirical models based on field-measured data (for sub-layer weighting and model validation), derived thermal comfort indices maps (PET in this case), and thermal zoning and design recommendation maps. To better support climate-responsive urban design for various climate conditions, more work is needed in the future. Firstly, nighttime and wintertime data need to be incorporated in empirical modeling, so that the urban morphological effect on microclimate and thermal comfort for difference time and seasons can be evaluated comprehensively. Secondly, the influence of the mesoscale climate such as regional wind information, based on mesoscale wind simulation or urban observatory databases, should be added as an additional sub-layer to providing better understating on urban ventilation.

It is necessary to discuss the robustness of the thermal atlas model, in particular the regression models built upon field data from limited measurement. Generally, more dependent variables are included in a multiple regression modeling, and more datasets from larger samples are needed. For the atlas model in this case study, we consider the sample size a reasonably good one for the parameters involved. Nevertheless, empirical models based on long-time continuous measurement would enhance the robustness. For this, practical issues related to logistics and management need to be handled with, when creating a network of urban sensors. Current advancement in smart city movement may see some opportunities, e.g., the smart street lampposts in Singapore, which to integrate WIFI, electric charging, public surveillance camera, and various sensors, e.g., pedestrian and vehicle motion, pollutant concentration, noise and light level, and weather conditions. Such lampposts with microclimate sensors installed, distributed within the district of interest, can be a key component of future urban thermal atlas system, in which the empirical model can be readily managed and updated, and citizens as well as urban managers be provided with latest visual microclimate information.

8.2 Lujiazui Elevated Walkway (LEW)

8.2.1 Introduction

Elevated walkways (EW) are an effective way to connect isolated buildings, enhance their accessibility, and vitalize commercial spaces at the elevated level. In high-density urban areas, carefully designed skywalk systems create a relatively pedestrian-friendly environment by distancing people from vehicle pollution and noise. Therefore, it has the potential to create safe and comfortable public space for social activities amidst busy urban centers.

Elevated walkways can be defined as "networks of above-grade connections between buildings that are often enclosed and climate controlled, and which link second-level corridors within buildings and various activity hubs, such as shops and offices" (Byers 1998) (pp. 11). In central urban areas, an EW system can facilitate pedestrian movement, improve accessibility to isolated urban buildings, protect pedestrians from vehicle pollution and noise, and provide shelter under adverse climate conditions, all contributing to a more pedestrian-friendly urban environment. There are some debates on whether or not EW systems will ruin street life in western cities (Rotmeyer 2006). In Asian cities, where population density is much higher, EW systems can greatly relieve the burden of crowded sidewalks on the ground levels. In some cases, it can help rebuild the pedestrian network, which will otherwise not work due to vehicle-oriented urban planning, for instance, the EW system in the Lujiazui CBD of Shanghai.

Famous examples of EW systems include the skyway in Minneapolis, Minnesota, USA and pedestrian skywalks in Calgary, Canada. These are both North American cities with cold climates featuring long and freezing winters. EW systems can also be found in cities under hot–humid subtropical climates. In contrary to the fully enclosed "tube" form in cold climates, EWs in warm and hot climates normally keep railings and overhangs where necessary for the sake of safety and protection, and open other surfaces to the ambient environment as much as possible, so as to enjoy natural ventilation while protecting pedestrians from summer sun and rain. Note that the EW system in some extreme climates (e.g., tropical climates) can be completely enclosed and fully air-conditioned. Pedestrian thermal comfort is well studied at the ground level. However, studies focusing on thermal comfort on an elevated level seem very limited. Compared with sidewalks at ground level, elevated walkways may be able to enjoy better ventilation but may also be exposed to more solar heat gain, and the composite effect is affected by surrounding urban geometry, fabric, and materials. It will be useful to investigate the relationship between variables of built environment and thermal comfort indices, so as to inform future EW design to achieve a more comfortable pedestrian environment. This study aims to investigate the microclimate and human thermal comfort of the Lujizaui Elevated Walkway (LEW) system.

The Lujiazui Elevated Walkway (LEW) is located in the Small Lujiazui CBD area. The purpose of introducing walkways at such a large scale is to improve the pedestrian environment for office commuters and tourists. The LEW comprises four parts, i.e., Oriental-Pearl Ring, Century Floating Pavilion, Century Sky Bridge, and Century Corridor (Fig. 8.9a). The length in total is 1373 m. The width ranges from 9.1 to 10.1 m (excluding enlarged plazas near subway entrance). Elevation is 8 m above ground. Construction is reinforced concrete and steel. The LEW connects all of the entrances of Lujiazui Station, Shanghai Metro Line 2, as well as five major large buildings: Super Brand Mall (a retail-recreational complex); Century floating pavilion (retail and restaurant); Shanghai International Financial Center (IFC) (retail and office); Jinmao Tower (retail and office); and Shanghai World Financial Center (SWFC) (retail and office).

Micrometeorological measurement was carried out in July 2014. Seven pairs of points were chosen, representing various scenarios of urban morphology on the

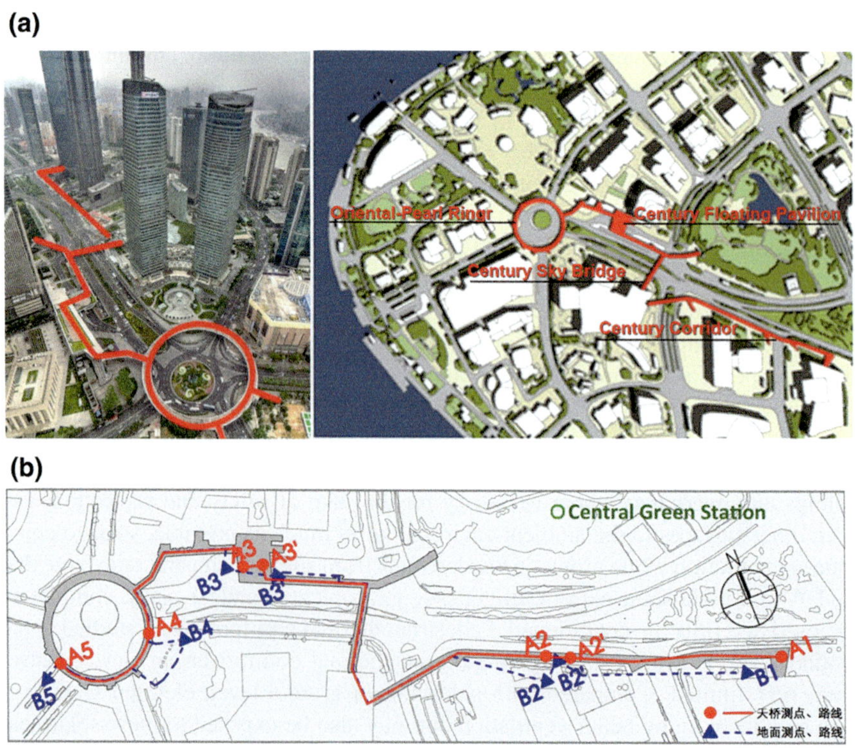

Fig. 8.9 Field measurement on the Lujiazui Elevated Walkway (LEW): measured points (**a**) and traverse routes (**b**)

walkway, and, on the sidewalk level, surface material, degree of space enclosure, green coverage, and degree of shading. Among them, four pairs of measurement points are selected to compare the effect of various shading devices on thermal comfort moderation. The two points in each pair are horizontally close to each other in order to control unmeasured effects of other thermal factors. Points with a prefix "A-" are located over the LEW, whereas points with a prefix "B-" are located under the LEW. These include A2 (in middle of the walkway and unshaded) versus A2′ (under a steel–glass constructed canopy), A3 (center of an elevated plaza near subway entrance) versus A3′ (the seat-and-rest area around the plaza, under a steel-glass constructed canopy), B2 (under a tree canopy) versus B2′ (directly under the LEW), and B3 (near the subway entrance, unshaded) versus B3′ (a small pedestrian rest area under and shaded by the LEW). Two pedestrian routes connect measurement points at the elevated level and at the ground level, respectively (Fig. 8.9b).

Four rounds of traverse measurements covering all points were carried out during four periods per day: 8–9:30 a.m., 10–11:30 a.m., 2:30–4:00 p.m., and 4:30–6:00 p.m., recording air temperature, relative humidity, wind velocity, and globe temperature using a portable microweather station. A reference station was set up on the open grass lawn of LZJ Central Green. During 8 a.m.–6 p.m., it continuously recorded global solar radiation and wind direction, in addition to the above-mentioned parameters. During the measurement period, the prevailing wind direction at the LJZ urban area is from the southeast (90–180 degree). The hourly mean wind velocity ranges from 0.7 to 1.0 m/s. Mean air temperature exceeded 30 °C even in the early morning (around 8:30 a.m.) and reached as high as 34 °C in the afternoon (between 3:00 and 3:30 p.m.).

8.2.2 Results

- Comparisons between, over, and under LEW

Inter-urban Temperature Differential (ITD) is the air temperature differential between measurement points and the Lujiazui Central Green Station (CGS). Wind velocity ratio (WVR) is the ratio of wind velocity at measurement points to that at the CGS.

Not surprisingly, ITD is higher at the points over LEW than those under LEW, due to less shading and thus more solar heat gain. However, it is counterintuitive to find that WVR over LEW is generally lower than that under LEW (Fig. 8.10).

The T_{mrt} values are all higher at the over-LEW points than their counterpoints under LEW, on the order of 2–6 °C, as are the PET values. The differences of PET are on the order of 1–3 °C (Fig. 8.11). According to (ISO-7726 1998), all of the points are hot (35–41 °C) during the measurement period, but clearly it was less uncomfortable under the LEW than over it.

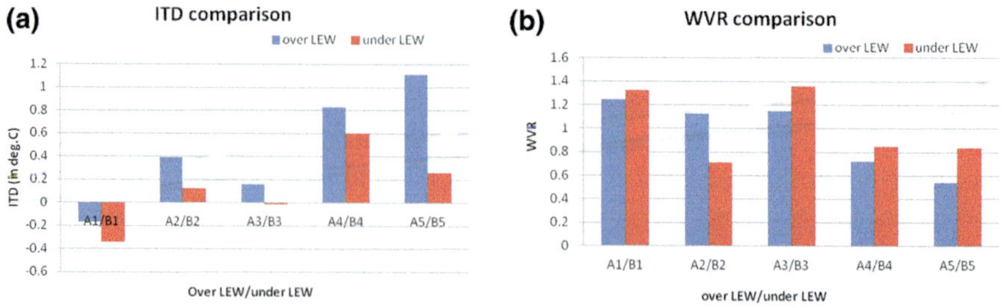

Fig. 8.10 Inter-urban Temperature Differential (ITD) (**a**) and Wind Velocity Ratio (WVR) (**b**) comparison below and above LEW

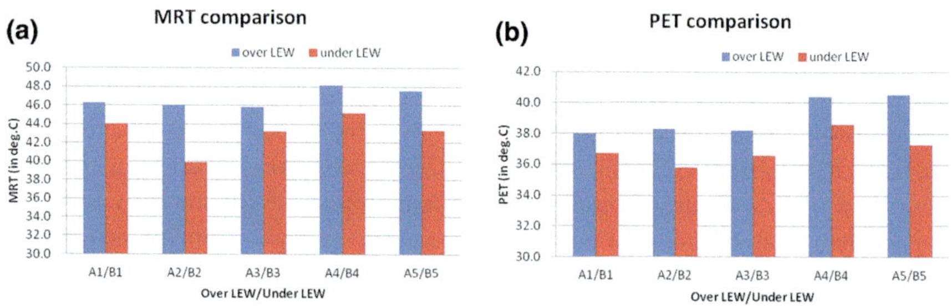

Fig. 8.11 Mean radiant temperature (T_{mrt}) (**a**) and physiological equivalent temperature (PET) (**b**) comparison below and above LEW

- Comparison between Shaded and UnShaded

As shown in Fig. 8.12, all shaded points showed lower ITD values, compared to the unshaded counterpart points. The differences range from 0.2 to 0.5 °C. The two shaded points at the ground level (B2′ and B3′) are markedly cooler than the reference Central Green station. Glass-shading demonstrated a clear effect on Ta reduction (A2′ and A3′), in the range of 0.1–0.3 °C, whereas the cooling effect by the LEW structure was higher (B2′ and B3′), on the order of 0.3–0.5 °C. The point under the LEW (B2′) was clearly cooler than the point under a tree (B2).

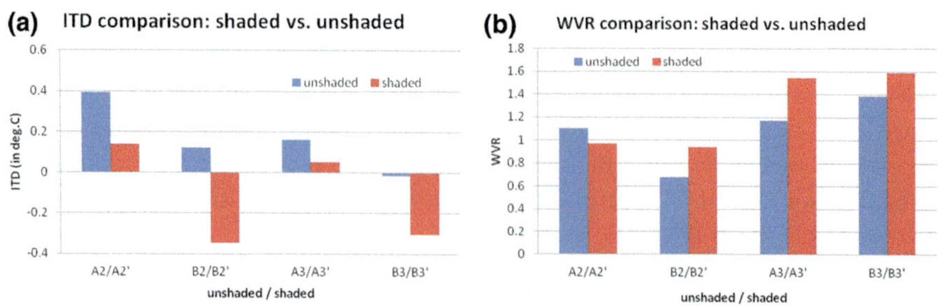

Fig. 8.12 ITD (**a**) and WVR (**b**) comparison shaded and unshaded

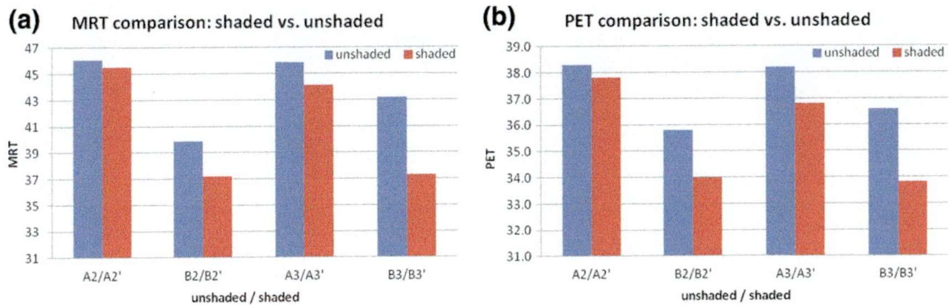

Fig. 8.13 T_{mrt} (**a**) and PET (**b**) comparison between shaded and unshaded

This is because a tree canopy, depending on the canopy geometry and leaf density, intercepts only a portion of incoming direct solar radiation, compared to the opaque structure of LEW. Regarding WVR, similar to the previous section, shaded points enjoyed higher WVR than their unshaded counterpart points, except for A2/A2′. As discussed previously, this can be caused by vertical thermal buoyancy and horizontal displacement ventilation between ground surfaces with different degrees of solar heating.

As shown in Fig. 8.13, the cooling effect of different shading devices becomes even clearer in T_{mrt} comparison: the glass–steel canopy showed limited T_{mrt} reduction, on the order of 0.5–1.5 °C. On the contrary, solid shading devices (elevated walkway in this case) lowered T_{mrt} by nearly 3 °C compared to a tree canopy shading (B2), and by about 6 °C compared to unshaded places (B3). The PET comparison has a similar pattern with T_{mrt}. The two points under the LEW (B2' and B3′) are classified as "warm" (29–35 °C in PET) while all other points are classified "hot" (35–41 °C in PET), including the tree shading point (B2) and two points under the semitransparent canopy (A2′ and A3′).

- Regression Analysis

Bivariate and multiple linear regression analyses are applied to identify the causal factors associated with temperature and thermal comfort indices. The significant level is set at 5%. SPSS software is used to carry out the statistical analysis on the traverse measurement data. The overall sample size including all of the traverse measurement points (see Fig. 8.9 for locations of the points) is 42. The dependent variable is air temperature (Ta), mean radiant temperature (T_{mrt}), and physiological equivalent temperature (PET). The independent variables include two point-specific variables, i.e., sky view factor (SVF) and green plot ratio (GPR), and one site-specific variable, i.e., background air temperature measured at CGS (Ta_cg).

Linear-fit estimates indicate that Ta is related to SVF with an R-square of 0.14, significant at the 0.05 level. A significant relationship exists between Ta and Ta_cg ($R^2 = 0.53$; Sig. level: 0.01) (Fig. 8.14). Air temperature variation is subject to many factors at different scales, and SVF alone cannot explain the major variation in air temperature. Although SVF as a crucial microscale parameter shows a statistically

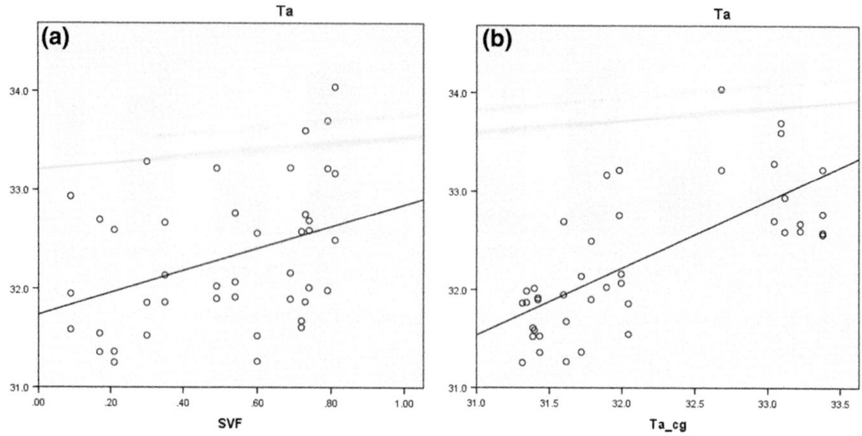

Fig. 8.14 Linear-fit estimates of air temperature (Ta) with **a** sky view factor (SVF) and **b** Ta_cg

significant relationship, its explanatory power is much less than the reference temperature recorded at the local scale, i.e., Ta_cg. Higher SVF tends toward increasing Ta, and a higher background temperature tends toward increasing Ta as well. Multiple regression incorporating SVF and Ta_cg yields the following equation. The model is capable of explaining about two-thirds of the variability in Ta (Eq. 8.4):

$$\mathrm{Ta} = 0.68 \cdot \mathrm{Ta\_cg} + 1.08 \cdot \mathrm{SVF} + 10.03\left(R^2 = 0.65,\ F = 35.5\right) \tag{8.4}$$

Linear-fit estimates indicate significant relationship of T_{mrt} with Ta_cg ($R^2 = 0.29$; Sig. level: 0.01), SVF ($R^2 = 0.36$; Sig. level: 0.01) and GPR ($R^2 = 0.28$; Sig. level: 0.01) (Fig. 8.15). Higher SVF tends toward increasing T_{mrt}, and higher background temperature tends toward increasing T_{mrt} as well, whereas higher greenery density tends toward lowering T_{mrt}. SVF shows a relatively low R-square value, due to the fact that T_{mrt} is highly dependent upon impinging solar radiation, and since SVF does not take solar geometry into account, it is not adequate to quantify solar radiation

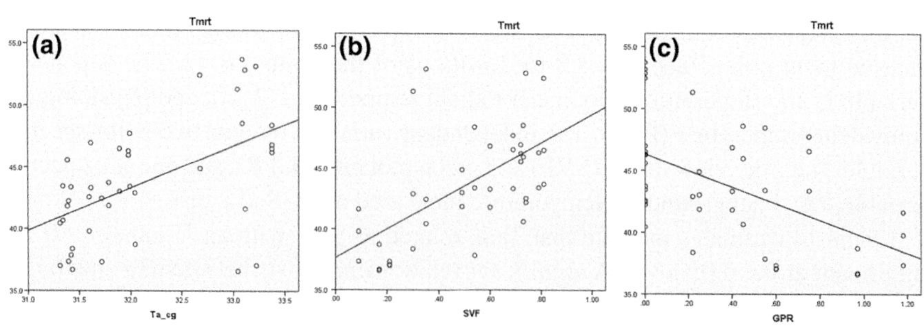

Fig. 8.15 Linear-fit estimates of mean radiant temperature (T_{mrt}) with **a** Ta_cg; **b** SVF and **c** Green plot ratio (GPR)

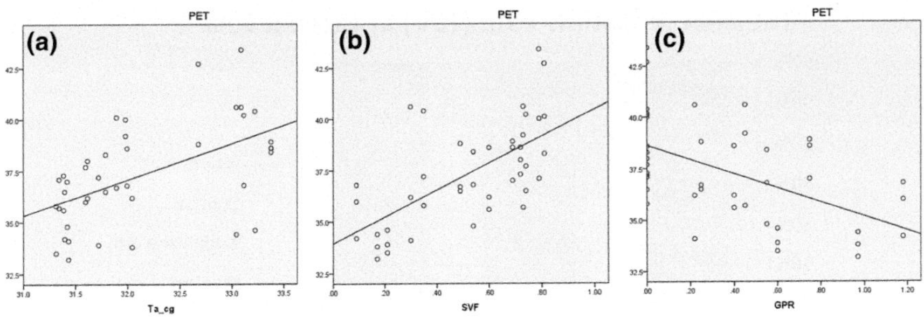

Fig. 8.16 Linear-fit estimates of PET with **a** Ta_cg; **b** SVF and **c** GPR

received at the location of interest. Greenery (trees, shrub, and grass) may modify T_{mrt} by tree canopy shading (direct solar radiation) and reducing ground albedo (reflected solar radiation) (Yang et al. 2015). Multiple regression incorporating SVF, GPR, and Ta_cg yields the following equation. The model is capable of explaining about 70% of the variability in T_{mrt} (Eq. 8.5):

$$T_{mrt} = 8.25 \cdot \text{SVF}—3.40 \cdot \text{GPR} + 3.42 \cdot \text{Ta\_cg}—69.02\left(R^2 = 0.69,\ F = 27.6\right) \tag{8.5}$$

Linear-fit estimates indicate a significant relationship of PET with Ta_cg ($R^2 = 0.29$; Sig. level: 0.01); SVF ($R^2 = 0.42$; Sig. level: 0.01) and GPR ($R^2 = 0.28$; Sig. level: 0.01) (Fig. 8.16). Higher SVF tends toward increasing PET, and higher background temperature tends toward increasing PET as well, whereas higher greenery density tends toward lowering PET.

Multiple regression incorporating SVF, GPR, and Ta_cg yields the following equation. The model is capable of explaining about 72% of the variability in PET (Eq. 8.6). The equation with standardized coefficients is as Eq. 8.7). Equation 8.8 is deducted from Eq. 8.2 of this chapter. By comparison, it can be seen that Eqs. 8.7 and 8.8 are similar in terms of variable composition and magnitudes of coefficients. Therefore, the robustness of the regression results is verified:

$$\text{PET} = 4.98 \cdot \text{SVF}—1.46 \cdot \text{GPR} + 1.73 \cdot \text{Ta\_cg}—20.33\ \left(R^2 = 0.72,\ F = 32.4\right) \tag{8.6}$$

$$\text{PET} = 0.49 \cdot \text{SVF}—0.23 \cdot \text{GPR} + 0.53 \cdot \text{Ta\_cg} \tag{8.7}$$

$$\text{PET} = 0.56 \cdot \text{SVF}—0.31 \cdot \text{GPR} + 0.38 \cdot \text{Ta\_cg}\left(R^2 = 0.76,\ F = 24\right) \tag{8.8}$$

- Questionnaire Survey

Overall, the respondents reported a warm-to-hot environment at LEW and at LCG (Fig. 8.17). The overall portion of respondents reporting warm to hot (+1 to +3) is about 77% above LEW and at LCG, whereas the percentage is about 8% lower in

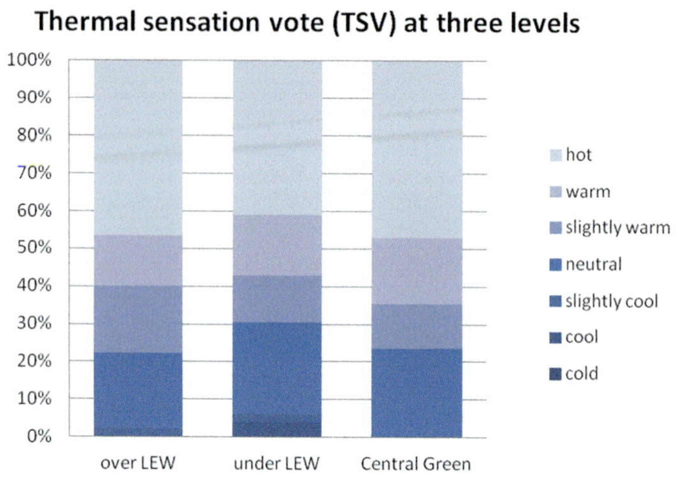

Fig. 8.17 Thermal sensation vote comparison

the group below LEW. About two-thirds of respondents above the LEW reported hot (+3), similar with the LCG reference station. In comparison, about 61% of the respondents below the LEW reported hot. In addition, 20–25% of the respondents reported neutral-to-cool at all three places. Note that all these responses were collected during the fourth round measurement (4:30–6 p.m.), when air temperature and solar radiation dropped down considerably compared to peak noon time.

More than 80% of respondents reported perceptible winds at all three places (Fig. 8.18). About 94% of respondents from above LEW reported perceptible winds

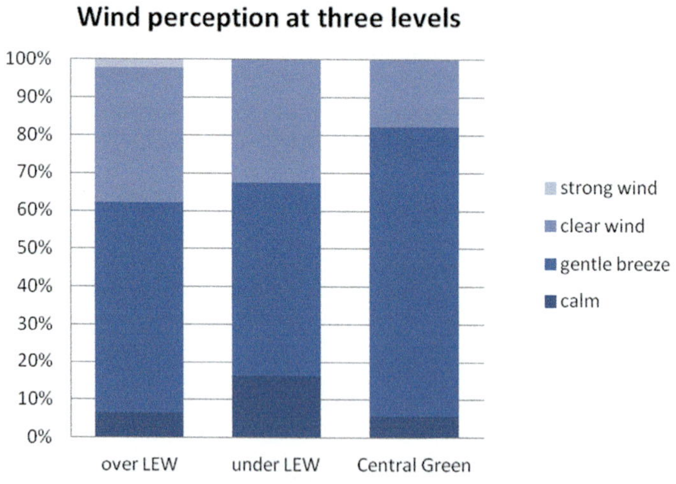

Fig. 8.18 Wind velocity perception comparison

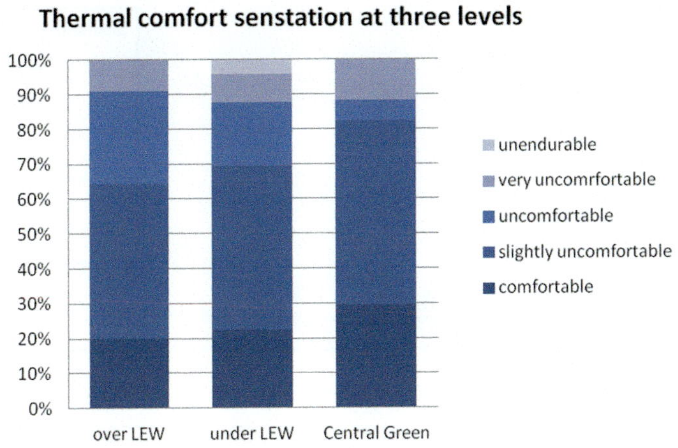

Fig. 8.19 Thermal comfort perception comparison

(+1 to +3, gentle breeze to strong wind), whereas about 84% from below LEW reported perceptible wind, 10% lower than that above LEW.

It is not surprising to find that respondents that felt comfortable comprise only around 20% both above and below the LEW (Fig. 8.19). More people felt comfortable at the under LEW level, but with only a marginal advantage of about 5%. About 30% reported being comfortable at the LCG.

8.2.3 *Discussion*

Climatically, sizable green spaces such as LCG can have a clear assimilating effect on the surrounding urbanized area, and results of the present study indicate that the degree of assimilation is proportional to the distance. In this study, the locations closer to the LCG were measured and found to have smaller Ta differences and higher velocity ratios than those farther away from it. Compared to the ground level, selected locations at the LEW level were measured with higher T_{mrt} on the order of 2–6 °C, higher ITD on the order of 0.2–0.8 °C, and lower WVR on the order of 0.1–0.3. The lower velocity ratio at higher elevation seems counterintuitive. A possible reason can be that the horizontal convection on the ground level was enhanced due to thermal buoyancy between shaded (directly under LEW) and unshaded places, i.e., thermal buoyancy causes uplift of warmer air at sun-lit spaces, and they were supplied by cooler air from surrounding shaded spaces (Fig. 8.20). Under hot and calm weather conditions, thermal buoyancy could be the major forces behind measured air movement at the pedestrian level (Yang et al. 2013). However, more data are to be collected before any solid conclusions can be made on this observation.

Increasing the height of EW could expose it to higher wind speed due to less ground friction. However, to achieve tangible improvement, the height may have to

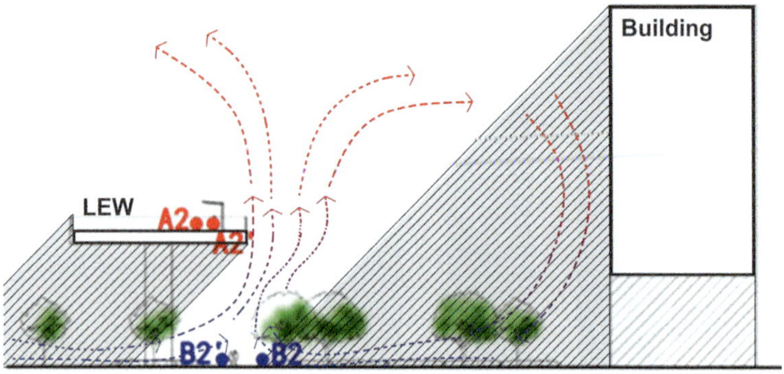

Fig. 8.20 Diagram showing possible thermal buoyancy circulation under LEW

be increased by a factor of two. This will significantly increase the cost of structure and lower the accessibility from ground level. Overall, comfort index PET indicates a hot thermal sensation for people (35–41 °C). However, PET was higher on the LEW level, on the order of 1–3 °C, indicating an even more uncomfortable thermal environment compared to the ground level.

Shading can be effective in reducing T_{mrt} and lowering PET, and thus can be en essential measure to improving thermal comfort. Field study further indicates that, among various materials, opaque shading with high thermal mass (concrete elevated walkway in this case) showed the best effect in lowering radiant temperature, on the order of 3–6 °C, followed by porous green mass (street tree canopy) (on the order of 1–3 °C), and semitransparent (tinted glass with steel frame) structure (on the order of 0.5–1.5 °C). Due to its high thermal mass, concrete surfaces maintain relatively lower surface temperature in addition to intercepting 100% direct solar radiation. Concrete shading is not new. Its application to building façades can be traced back to the brise-soleil populated by Le Corbusier; famous examples include Chandigarh City Hall in India and Unité d'habitation in France. However, aesthetically, its raw and "brutalism" look might seem incompatible with the modern glass–steel towers commonly found in CBD areas. Alternatively, various shading devices that are lightweight and opaque can be applied in the EW design. Note that vegetation density is found significantly correlated with mean radiant temperature and the thermal comfort index. At the LEW level, planting trees would be structurally difficult and not cost-effective. The effective strategy to increase greenery mass can be shading canopy by climbing plants (Fig. 8.21).

Under the peak summer weather conditions in Shanghai, outdoor thermal comfort cannot be met even with sufficient shading. A previous study shows that when outdoor air temperature is in the range of 30–32 °C, shaded street space can be comfortable with wind velocity on the order of 2.2–3.6 m/s. The measured WVR and reference WV indicate that, on average, this WV range was not achieved during the field measurement. To boost air movement at the pedestrian level over the LEW, electrical fans can be installed on the shading canopy. For instance, in the Clarke

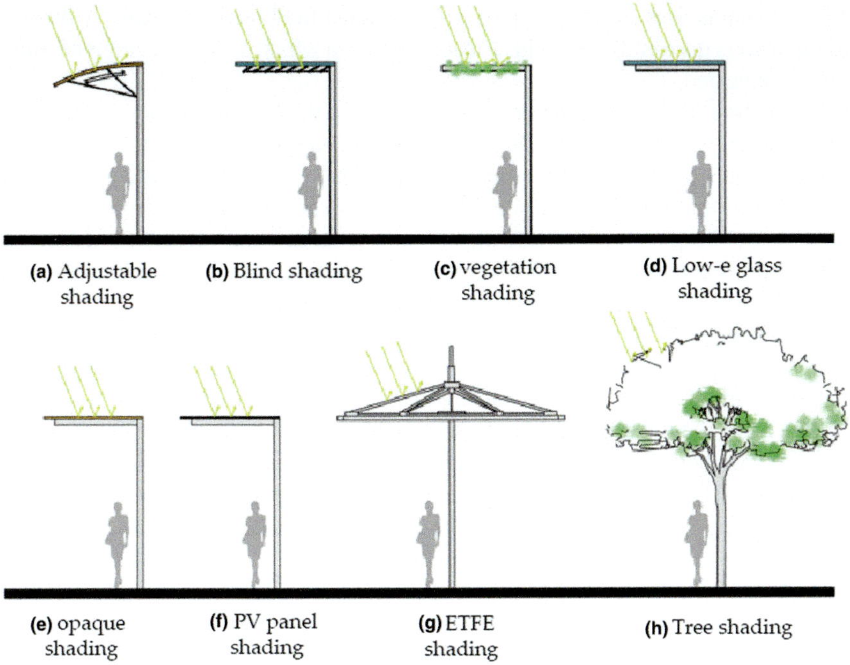

Fig. 8.21 Diagrams of various forms of shading form for elevated walkway (EW)

Quay redevelopment in Singapore, air ducts and mechanical fans were incorporated into the canopy structure over the pedestrian area to promote air ventilation under nearly calm weather conditions. Combined with water misting devices to lower sensible heat (by increasing latent heat), the LEW canopy can be upgraded into a "cool corridor" during the daytime. The installation and maintenance can be costly, but for a high-profile walkway being heavily used such as LEW, it can be worthwhile to invest.

8.3 Conclusion

This first case study presents a thermal atlas system developed for supporting climate-responsive urban design under the subtropical climate of Shanghai. The system is based on empirical modeling and GIS-based spatial analysis by DEM modeling. Carefully selected morphological variables and control variables have been shown to be able to explain nearly 80% of the variability in the thermal comfort index of PET. The morphological variables were then calculated and mapped in the virtual urban space, and a thermal atlas was generated by overlaying the key morphological variable layers based on the coefficients determined by the empirical models. A set of PET maps is thus generated and verified against field-measured data. A thermal zoning map is derived from the PET maps, based on which design recommendations can be suggested.

To our best knowledge, the present study is the first attempt to combine empirical modeling with GIS-based spatial analysis techniques, in order to improve the accuracy and validity of urban climate mapping to suit the microscale application. It is believed that this framework maintains the advantages in data visualization of previous UC-Map methods, while improves its accuracy and validity for urban and building design implication through empirical modeling and weighting, and thus has a good potential in climate-responsive design research and practice.

The second case study is supplemented by an investigation on the relationship between urban morphology and urban microclimate and thermal comfort of the recently completed Lujiazui Elevated Walkway (LEW) system. The data analysis indicates that: The measured locations over the LEW are thermally more uncomfortable than those below it. Air temperature was higher, whereas wind velocity is lower on the LEW level than on the ground level, which is counterintuitive. It is possible that the horizontal convection on the ground level was enhanced due to thermal buoyancy between shaded and unshaded places.

Indicated by the calculated thermal comfort index PET, it was averagely hot both over and below the LEW during the measured period, although PET was 1–3 °C lower at below the LEW. In addition, about 80% of respondents reported being uncomfortable above the LEW, whereas this was 5% lower at below the LEW. Shaded locations can be warm while unshaded places can be hot indicated by PET. Opaque concrete shading is most effective in lowering T_{mrt}, followed by tree canopy and glass–steel canopy. To achieve a thermally comfortable LEW, passive cooling systems such as shading are vital but not enough. Active energy measures can be combined with shading devices, to increase air movement and reduce sensible heat, by a carefully integrated system design.

References

Byers JP (1998) Breaking the ground plane: the evolution of grade separated cities in North America. Department of Geography, University of Minnesota, Minneapolis, MN

Chen Q (2004) Using computational tools to factor wind into architectural environment design. Energy Build 36:1197–1209

Chen L, Ng E (2011) Quantitative urban climate mapping based on a geographical database: a simulation approach using Hong Kong as a case study. Int J Appl Earth Obs Geoinf 13:586–594

Chen L, Ng E, An X, Ren C, Lee M, Wang U, He Z (2012) Sky view factor analysis of street canyons and its implications for daytime intra-urban air temperature differentials in high-rise, high-density urban areas of Hong Kong: a GIS-based simulation approach. Int J Climatol 32:121–136

Hii JC, Wong NH, Jusuf SK (2015) Anthropogenic heat contribution to air temperature increase at pedestrian height in Singapore's high density Central Business District (CBD). In: 9th international conference on urban climate (ICUC9). Toulouse France. 20th-24th July 2015, pp 1–6

Hoppe P (1999) The physiological equivalent temperature—a universal index for the biometeorological assessment of the thermal environment. Int J Biometeorol 43:71–75

ISO-7726 (1998) Ergonomics of the thermal environment—Instruments for measuring physical quantities. ISO, Geneva

Lin T-P, Matzarakis A, Hwang R-L (2010) Shading effect on long-term outdoor thermal comfort. Build Environ 45:213–221

Matzarakis A, Mayer H, Iziomon MG (1999) Applications of a universal thermal index: physiological equivalent temperature. Int J Biometeorol 43:76–84

Ng E (2012) Towards planning and practical understanding of the need for meteorological and climatic information in the design of high-density cities: a case-based study of Hong Kong. Int J Climatol 32:582–598

Niu J, Liu J, Lee T-C, Lin Z, Mak C, Tse K-T, Tang B-S, Kwok KCS (2015) A new method to assess spatial variations of outdoor thermal comfort: onsite monitoring results and implications for precinct planning. Build Environ 91:263–270

Planning Department Hong Kong SAR (2009) Urban climatic map and standards for wind environment—feasibility study—final Report. Department of Architecture CUHK, Hong Kong

Rotmeyer J (2006) Can elevated pedestrian walkways be sustainable? WIT Trans Ecol Environ 93:293–302

Tokyo Metropolitan Government TMG (2005) The thermal environment map and areas designated for the implementation of measures against the Heat Island Phenomenon. In: Bureau of environment, Bureau of urban development, Tokyo Metropolitan Government (TMG), Tokyo, Japan

VDI. (1997) VDI-Guideline 3787, Part 1: environmental meteorology-climate and air pollution maps for cities and regions. Beuth Verlag, Berlin

Yang F, Chen L (2016) Developing a thermal atlas for climate-responsive urban design based on empirical modeling and urban morphological analysis. Energy Buildings, 111:120–130. https://doi.org/10.1016/j.enbuild.2015.11.047

Yang F, Lau S, Qian F (2015) Cooling performance of residential greenery in localised urban climates: a case study in Shanghai China. Int J Environ Technol Manag 18:478–503

Yang F, Qian F, Lau SSY (2013) Urban form and density as indicators for summertime outdoor ventilation potential: a case study on high-rise housing in shanghai. Build Environ 70:122–137

Yang F, Qian F, Zhao W (2016) Towards a climate-responsive vertical pedestrian system: an empirical study on an elevated walkway in Shanghai China. Sustainability 8:744–758. https://doi.org/10.3390/su8080744

Printed by Printforce, the Netherlands